Cracking the

AP®

CHEMISTRY EXAM

Revised 2014 Edition

Paul Foglino, Nicolas Leonardi,
and The Staff of the Princeton Review

PrincetonReview.com

Random House, Inc. New York

The Princeton Review, Inc.
111 Speen Street, Suite 550
Framingham, MA 01701
E-mail: editorialsupport@review.com

Published in the United States by Random House LLC, New York, and
simultaneously in Canada by Random House of Canada Limited, Toronto.
A Penguin Random House Company.

ISBN: 978-0-8041-2475-1
ISSN: 1092-0102

AP and Advanced Placement Program are registered trademarks of the
College Board, which does not sponsor or endorse this product.

The Princeton Review is not affiliated with Princeton University.

Editor: Calvin S. Cato
Production Editor: Harmony Quiroz
Production Artists: Deborah Silvestrini

Printed in the United States of America on partially recycled paper.

10 9 8 7 6 5 4 3 2 1

Revised 2014 Edition

Editorial

Rob Franek, Senior VP, Publisher
Mary Beth Garrick, Director of Production
Selena Coppock, Senior Editor
Calvin Cato, Editor
Kristen O'Toole, Editor
Meave Shelton, Editor
Alyssa Wolff, Editorial Assistant

Random House Publishing Team

Tom Russell, Publisher
Alison Stoltzfus, Publishing Manager
Dawn Ryan, Associate Managing Editor
Ellen Reed, Production Manager
Erika Pepe, Associate Production Manager
Kristin Lindner, Production Supervisor
Andrea Lau, Designer

Acknowledgments

The author would like to thank John Katzman for his oversight on this project and Rebecca Lessem for her work editing the first edition of this title. Thanks to Rachel Warren, whose guidance and tireless effort helped to convert chemistry to English. Thanks to Eric Payne for his rigorous attention to detail. Thanks to Tom Meltzer for his advice and Libby O'Connor for her patience. Additional thanks are in order to Robbie Korin, James Karb, and Chris Volpe.

The Princeton Review would like to give special thanks to Nicolas Leonardi for his hard work in updating AP Chemistry to in preparation for the changes to the 2014 edition of the AP Chemistry Exam.

About the Author

Paul Foglino has taught for The Princeton Review for more than a decade. He has written and edited course materials for the SAT, GRE, GMAT, and MCAT courses. He is author of *Math Smart Jr. II* and coauthor of *Cracking the CLEP*. Foglino studied English and electrical engineering at Columbia University, but he remains convinced that he learned everything he ever needed to know in junior high school.

Contents

Part I
Using This Book to Improve Your AP Score

PREVIEW ACTIVITY: YOUR KNOWLEDGE, YOUR EXPECTATIONS

Your route to a high score on the AP Chemistry Exam depends a lot on how you plan to use this book. Respond to the following questions.

1. Rate your level of confidence about your knowledge of the content tested by the Chemistry AP Exam:

    A. Very confident—I know it all
    B. I'm pretty confident, but there are topics for which I could use help
    C. Not confident—I need quite a bit of support
    D. I'm not sure

2. Circle your goal score for the AP Chemistry Exam.

    5   4   3   2   1   I'm not sure yet

3. What do you expect to learn from this book? Circle all that apply to you.

    A. A general overview of the test and what to expect
    B. Strategies for how to approach the test
    C. The content tested by this exam
    D. I'm not sure yet

YOUR GUIDE TO USING THIS BOOK

This book is organized to provide as much—or as little—support as you need, so you can use this book in whatever way will be most helpful to improving your score on the AP Chemistry Exam.

- The remainder of **Part One** will provide guidance on how to use this book and help you determine your strengths and weaknesses

- **Part Two** of this book will:
    - provide information about the structure, scoring, and content of the Chemistry Exam
    - help you to make a study plan
    - point you towards additional resources

- **Part Three** of this book will explore various strategies:
    - how to attack multiple choice questions
    - how to write high scoring constructed response answers
    - how to manage your time to maximize the number of points available to you

- **Part Four** of this book covers the content you need for your exam.

- **Part Five** of this book contains practice tests.

You may choose the use some parts of this book over others, or you may work through the entire book. This will depend on your needs and how much time you have. Let's now look how to make this determination.

HOW TO BEGIN

1. Take a Test

   Before you can decide how to use this book, you need to take a practice test. Doing so will give you insight into your strengths and weaknesses, and the test will also help you make an effective study plan. If you're feeling test-phobic, remind yourself that a practice test is a tool for diagnosing yourself—it's not how well you do that matters but how you use information gleaned from your performance to guide your preparation.

   So, before you read further, take AP Chemistry Practice Test 1 starting at page 259 of this book. Be sure to do so in one sitting, following the instructions that appear before the test.

2. Check Your Answers

   Using the answer key on page 285, count how many multiple choice questions you got right and how many you missed. Don't worry about the explanations for now, and don't worry about why you missed questions. We'll get to that soon.

3. Reflect on the Test

   After you take your first test, respond to the following questions:

   - How much time did you spend on the multiple choice questions?

   - How much time did you spend on each long form constructed response question? What about each short form constructed response question?

   - How many multiple choice questions did you miss?

   - Do you feel you had the knowledge to address the subject matter of the constructed response questions?

   - Do you feel you wrote well organized, thoughtful answers to the constructed response questions?

- Circle the content areas that were most challenging for you and draw a line through the ones in which you felt confident/did well.

  - **Big Idea #1:** Atoms, Elements, and the Building Blocks of Matter

  - **Big Idea #2:** Chemical and Physical Properties of Matter

  - **Big Idea #3:** Chemical Reactions, Energy Changes, and Redox Reactions

  - **Big Idea #4:** Chemical Reactions and their Rates

  - **Big Idea #5:** Laws of Thermodynamics and Changes in Matter

  - **Big Idea #6:** Equilibrium, Acids and Bases, Titrations and Solubility

4. **Read Part Two of this Book and Complete the Self-Evaluation**

As discussed in the Goals section above, Part Two will provide information on how the test is structured and scored. It will also set out areas of content that are tested.

As you read Part Two, re-evaluate your answers to the questions above. At the end of Part Two, you will revisit and refine the questions you answer above. You will then be able to make a study plan, based on your needs and time available, that will allow you to use this book most effectively.

5. **Engage with Parts Three and Four as Needed**

Notice the word *engage*. You'll get more out of this book if you use it intentionally than if you read it passively, hoping for an improved score through osmosis.

Strategy chapters will help you think about your approach to the question types on this exam. Part Three will open with a reminder to think about how you approach questions now and then close with a reflection section asking you to think about how/whether you will change your approach in the future.

Content chapters are designed to provide a review of the content tested on the AP Chemistry Exam, including the level of detail you need to know and how the content is tested. You will have the opportunity to assess your mastery of the content of each chapter through test-appropriate questions and a reflection section.

6. **Take Test 2 and Assess Your Performance**

Once you feel you have developed the strategies you need and gained the knowledge you lacked, you should take Test 2, which starts at page 321 of this book. You should do so in one sitting, following the instructions at the beginning of the test.

When you are done, check your answers to the multiple choice sections. See if a teacher will read your answers to the constructed response questions and provide feedback.

Once you have taken the test, reflect on what areas you still need to work on, and revisit the chapters in this book that address those deficiencies. Through this type of reflection and engagement, you will continue to improve.

7. **Keep Working**

After you have revisited certain chapters in this book, continue the process of testing, reflection, and engaging with the second test. Each time, consider what additional work you need to do and how you will change your strategic approach to different parts of the test.

Part II
About the
AP Chemistry
Exam

THE STRUCTURE OF THE AP CHEMISTRY EXAM

The AP Chemistry Exam is a three-hour-long, two-section test that attempts to cover the material you would learn in a college first-year chemistry course. The first section is a 90-minute 60 question multiple-choice section. The second section is also 90 minutes and consists of 3 long-form constructed response questions and 4 short-form constructed response questions.

The multiple choice section is scored by a computer and the constructed response questions are scored by a committee of high school and college teachers. The constructed response questions are graded according to a standard set at the beginning of the grading period by the chief faculty consultants. Inevitably, the grading of Section II is never as consistent or accurate as the grading of Section I.

The AP Chemistry exam is changing for the 2014 admission, and as such there is no data on the necessary raw score to get a 3, 4, or 5. The breakdown will not be decided until after the exams have been scored, and with no past exams using this curriculum there is no existing standard to judge yourself by. Just do your best!

OVERVIEW OF CONTENT TOPICS

The concepts of the AP Chemistry exam are broken down into six major themes defined by the College Board as the Big Ideas. Rather than learning multiple disparate topics (as has been done in the past), these Big Ideas interconnect principles within these topics that describe fundamental chemical phenomena. The six Big Ideas are as follows:

- Big Idea #1: The chemical elements are fundamental building materials of matter, and all matter can be understood in terms of arrangements of atoms. These atoms retrain their identity in chemical reactions.

- Big Idea #2: Chemical and physical properties of materials can be explained by the structure and the arrangement of atoms, ions, or molecules and the forces between them.

- Big Idea #3: Changes in matter involve the rearrangement and/or reorganization of atoms and/or the transfer of electrons.

- Big Idea #4: Rates of chemical reactions are determined by details of the molecular collisions.

- Big Idea #5: The laws of thermodynamics describe the essential role of energy and explain and predict the direction of changes in matter.

- Big Idea #6: Any bond or intermolecular attraction that can be formed can by broken. These two processes are in a dynamic competition, sensitive to initial conditions and external perturbations.

This book has been arranged to teach chemistry topics grouped to each of these Big Ideas.

HOW AP EXAMS ARE USED

Different colleges use AP Exam scores in different ways, so it is important that you go to a particular college's web site to determine how it uses AP Exam scores. The three items below represent the main ways in which AP Exam scores can be used:

- **College Credit.** Some colleges will give you college credit if you score well on an AP Exam. These credits count towards your graduation requirements, meaning that you can take fewer courses while in college. Given the cost of college, this could be quite a benefit, indeed.

- **Satisfy Requirements.** Some colleges will allow you to "place out" of certain requirements if you do well on an AP Exam, even if they do not give you actual college credits. For example, you might not need to take an introductory-level course, or perhaps you might not need to take a class in a certain discipline at all.

- **Admissions Plus.** Even if your AP Exam will not result in college credit or even allow you to place out of certain courses, most colleges will respect your decision to push yourself by taking an AP Course or even an AP Exam outside of a course. A high score on an AP Exam shows mastery of more difficult content than is taught in many high school courses, and colleges may take that into account during the admissions process.

OTHER RESOURCES

There are many resources available to help you improve your score on the Chemistry Exam, not the least of which are your **teachers.** If you are taking an AP class, you may be able to get extra attention from your teacher, such as obtaining feedback on your constructed response questions. If you are not in an AP course, reach out to a teacher who teaches Chemistry, and ask if the teacher will review your constructed response questions or otherwise help you with content.

Another wonderful resource is **AP Central,** the official site of the AP Exams. The scope of the information at this site is quite broad and includes:

- Course Description, which includes details on what content is covered and sample questions

- Sample questions for the new May 2014 exam

- Constructed response question prompts and multiple choice questions from previous years

The AP Central home page address is: **http://apcentral.collegeboard.com/apc/ Controller.jpf.**

For up-to-date information about the ongoing changes to the AP Chemistry Exam Course, please visit: **http://advancesinap.collegeboard.org/science/chemistry.**

Finally, **The Princeton Review** offers tutoring and small group instruction. Our expert instructors can help you refine your strategic approach and add to your content knowledge. For more information, call 1-800-2REVIEW.

DESIGNING YOUR STUDY PLAN

As part of the Introduction, you identified some areas of potential improvement. Let's now delve further into your performance on Test 1, with the goal of developing a study plan appropriate to your needs and time commitment.

Read the answers and explanations associated with the Multiple Choice questions (starting at page 286). After you have done so, respond to the following questions:

- Review the Overview of Content Topics at page 8. Next to each topic, indicate your rank of the topic as follows: "1" means "I need a lot of work on this," "2" means "I need to beef up my knowledge," and "3" means "I know this topic well."

- How many days/weeks/months away is your exam?

- What time of day is your best, most focused study time?

- How much time per day/week/month will you devote to preparing for your exam?

- When will you do this preparation? (Be as specific as possible: Mondays & Wednesdays from 3 to 4 pm, for example)

- Based on the answers above, will you focus on strategy (Part Two) or content (Part Three) or both?

- What are your overall goals in using this book?

Part III
Test-Taking Strategies for the AP Chemistry Exam

Chapter 1
How to Approach Multiple-Choice Questions

THE BASICS

Section I of the test is composed of 60 multiple-choice questions, for which you are allotted 90 minutes. This part is worth 50 percent of your total score.

For this section, you will be given a periodic table of the elements along with a sheet that lists common chemistry formulas, and you may NOT use a calculator. The College Board says that this is because the new scien- tific calculators not only program and graph but also store information—and they are afraid you'll use this function to cheat!

On the multiple-choice section, you receive 1 point for a correct answer. There is no penalty for leaving a question blank or getting a question wrong.

PACING

According to the College Board, the multiple choice section of the AP Chemistry Exam covers more material than any individual student is expected to know. Nobody is expected to get a perfect or even near perfect score. What does that mean to you?

Use the Two-Pass System

Go through the multiple-choice section twice. The first time, do all the questions that you can get answers to immediately. That is, do the questions with little or no math and questions on chemistry topics in which you are well versed. Skip questions on topics that make you uncomfortable. Also, you want to skip the ones that look like number crunchers (even without a calculator, you may still be expected to crunch a few numbers). Circle the questions that you skip in your test booklet so you can find them easily during the second pass. Once you've done all the questions that come easily to you, go back and pick out the tough ones that you have the best shot at.

In general, the questions near the end of the section are tougher than the questions near the beginning. You should keep that in mind, but be aware that each person's experience will be different. If you can do acid-base questions in your sleep, but you'd rather have your teeth drilled than draw a Lewis diagram, you may find questions near the end of the section easier than questions near the beginning.

That's why the Two-Pass System is so handy. By using it, you make sure you get to see all the questions you can get right, instead of running out of time because you got bogged down on questions you couldn't do earlier in the test.

This brings us to another important point.

Don't Turn a Question into a Crusade!

Most people don't run out of time on standardized tests because they work too slowly. Instead, they run out of time because they spend half the test wrestling with two or three particular questions.

You should never spend more than a minute or two on any question. If a question doesn't involve calculation, then either you know the answer, you can make an educated guess, or you don't know the answer. Figure out where you stand on a question, make a decision, and move on.

Any question that requires more than two minutes worth of calculations probably isn't worth doing. Remember: Skipping a question early in the section is a good thing if it means that you'll have time to get two correct answers later on.

GUESSING

You get one point for every correct answer on the multiple-choice section. Guessing randomly neither helps you nor hurts you. Educated guessing, however, will help you.

Use Process of Elimination (POE) to Find Wrong Answers

There is a fundamental weakness to a multiple-choice test. The test makers must show you the right answer, along with three wrong answers. Sometimes seeing the right answer is all you need. Other times you may not know the right answer, but you may be able to identify one or two of the answers that are clearly wrong. Here is where you should use POE to take an educated guess.

Look at this hypothetical question.

1. Which of the following compounds will produce a purple solution when added to water?
   (A) Brobogdium rabelide
   (B) Diblythium perjuvenide
   (C) Sodium chloride
   (D) Carbon dioxide

You should have no idea what the correct answer is because three of these compounds are made up, but you do know something about the obviously wrong answers. You know that sodium chloride, choice (C), and carbon dioxide, choice (D), do not turn water purple. So, using POE, you have a 50 percent chance at guessing the correct answer. Now the odds are in your favor. Now you should guess.

Guess and Move On

Remember that you're guessing. Pondering the possible differences between bro-bogdium rabelide and diblythium perjuvenide is a waste of time. Once you've taken POE as far as it will go, pick your favorite letter and move on.

The multiple-choice section is the exact opposite of the free-response section. It's scored by a machine. There's no partial credit. The computer doesn't know, or care if you know, why an answer is correct. All the computer cares about is whether you blackened in the correct oval on your score sheet. You get the same number of points for picking (B) because you know (A) is wrong and (B) is a nicer letter than (C) or (D) as you would for picking (B) because you fully understood the subtleties of an electrochemical process.

ABOUT CALCULATORS

You will NOT be allowed to use a calculator on this section. That shouldn't worry you. All it means is that there won't be any questions in the section that you'll need a calculator to solve.

Most of the calculation problems will have fairly user-friendly numbers—that is, numbers with only a couple of significant digits, or things like "11.2 liters of gas at STP" or "160 grams of oxygen" or "a temperature increase from 27°C to 127°C." Sometimes these user-friendly numbers will actually point you toward the proper steps to take in your calculations.

Don't be afraid to make rough estimates as you do your calculations. Sometimes knowing that an answer is closer to 50 than to 500 will enable you to pick the correct answer on a multiple-choice test (if the answer choices are far enough apart). Once again, the rule against calculators works in your favor because the College Board will not expect you to do very precise calculations by hand.

There may be a couple of real number-crunching problems on the test. If you can recognize them quickly, these are good ones to skip. There's no point in spending five minutes crunching numbers to get one problem right if that time could be better used in getting three others correct later in the test.

REFLECT

Respond to the following questions:

- How long will you spend on multiple-choice questions?

- How will you change your approach to multiple-choice questions?

- What is your multiple-choice guessing strategy?

Chapter 2
How to Approach
Constructed
Response Questions

CRACKING THE CONSTRUCTED RESPONSE SECTION

Section II is composed of a series of seven constructed response questions, all of which are required. You will be allotted 90 minutes to complete this section, which is worth 50 percent of your total score.

Cracking the Math Problems

You want to show the graders that you can do chemistry math, so here are some suggestions.

Show every step of your calculations on paper

This section is the opposite of multiple choice. You don't just get full credit for writing the correct answer. You get most of your points on this section for showing the process that got you to the answer. The graders give you partial credit when you show them that you know what you're doing. So even if you can do a calculation in your head, you should set it up and show it on the page.

By showing every step, or explaining what you're doing in words, you insure that you'll get all the partial credit possible, even if you screw up a calculation.

Include units in all your calculations

Scientists like units in calculations. Units make scientists feel secure. You'll get points for including them and you may lose points for leaving them out.

Remember significant figures

You can lose one point per question if your answer is off by more than one significant figure. Without getting too bent out of shape about it, try to remember that a calculation is only as accurate as the least accurate number in it.

The graders will follow your reasoning, even if you've made a mistake

Often, you are asked to use the result of a previous part of a problem in a later part. If you got the wrong answer in part (a) and used it in part (c), you can still get full credit for part (c), as long as your work is correct based on the number that you used. That's important, because it means that botching the first part of a question doesn't necessarily sink the whole question.

Remember the mean!

So let's say that you could complete only parts (a) and (b) on the required equilibrium problem. That's 4 or 5 points out of 10, tops. Are you doomed? Of course not. You're above average. If this test is hard on you, it's probably just as hard on everybody else. Remember: You don't need anywhere near a perfect score to get a 5, and you can leave half the test blank and still get a 4!

STRATEGIES FOR CRACKING THE CONSTRUCTED RESPONSE SECTION

This section is here to test whether you can translate chemistry into English. The term "essay" is a little misleading because all of these questions can be answered in two or three simple sentences, or with a simple diagram or two. Here are some tips for answering the seven essay questions on Part II.

Show that you understand the terms used in the question

If they ask you why sodium and potassium have differing first ionization energies, the first thing you should do is tell them what ionization energy is. That's probably worth the first point of partial credit. Then you should tell them how the differing structures of the atoms make for differing ionization energies. That leads to the next tip.

Take a step-by-step approach

Grading these tests is hard work. Breaking a question into parts in this way makes it easier on the grader, who must match your response to a set of guidelines he or she has been given that describe how to assign partial and full credit.

Each grader scores each test based on these rough guidelines that are established at the beginning of the grading period. For instance, if a grader has 3 points for the question about ionization energies, the points might be distributed the following way:

- One point for understanding ionization energy.
- One point for explaining the structural difference between sodium and potassium.
- One point for showing how this difference affects the ionization energy.

You can get all three points for this question if the grader thinks that all three concepts are addressed implicitly in your answer, but by taking a step by step approach, you improve your chances of explicitly addressing the things that a grader has been instructed to look for. Once again, grading these tests is hard work; graders won't know for sure if you understand something unless you tell them.

This leads us to an obvious point.

Write neatly

Even if writing neatly means working at half-speed. You can't get points for answers if the graders can't understand them. Of course, this applies to the rest of the constructed response section as well.

The graders will follow your reasoning, even if you've made a mistake

Just like in the problems section, you might be asked to use the result of a previous part of a problem in a later part. If you decide (incorrectly) that an endothermic reaction in part (a) is exothermic, you can still get full credit in part (c) for your wrong answer about the reaction's spontaneity, as long as your answer in (c) is correct based on an exothermic reaction.

REFLECT

Respond to the following questions:

- How much time will you spend on the short constructed response questions? What about the long constructed response questions?

- What will you do before you begin writing your constructed response answers?

- Will you seek further help, outside of this book (such as a teacher, tutor, or AP Central), on how to approach the questions that you will see on the AP Chemistry exam?

Part IV
Content Review for the AP Chemistry Exam

Chapter 3
Big Idea #1: Atoms, Elements, and the Building Blocks of Matter

The chemical elements are fundamental building materials of matter, and all matter can be understood in terms of arrangements of atoms. These atoms retain their identity in chemical reactions.

THE PERIODIC TABLE

The most important tool you will use on this test is the Periodic Table of the Elements.

PERIODIC TABLE OF THE ELEMENTS

1 IA																	18 VIIIA
1 **H** 1.008	2 IIA											13 IIIA	14 IVA	15 VA	16 VIA	17 VIIA	2 **He** 4.00
3 **Li** 6.94	4 **Be** 9.01											5 **B** 10.81	6 **C** 12.01	7 **N** 14.1	8 **O** 16.00	9 **F** 19.00	10 **NE** 20.18
11 **Na** 22.99	12 **Mg** 24.30	3 IIIB	4 IVB	5 VB	6 VIB	7 VIIB	8	9 VIIIB	10	11 IB	12 IIB	13 **Al** 26.98	14 **Si** 28.09	15 **P** 30.97	16 **S** 32.06	17 **Cl** 35.45	18 **Ar** 39.95
19 **K** 39.10	20 **Ca** 40.08	21 **Sc** 44.96	22 **Ti** 47.90	23 **V** 50.94	24 **Cr** 52.00	25 **Mn** 59.94	26 **Fe** 55.85	27 **Co** 58.93	28 **Ni** 58.69	29 **Cu** 63.55	30 **Zn** 65.39	31 **Ga** 69.72	32 **Ge** 72.59	33 **As** 74.92	34 **Se** 78.96	35 **Br** 79.90	36 **Kr** 83.80
37 **Rb** 85.47	38 **Sr** 87.62	39 **Y** 88.91	40 **Zr** 91.22	41 **Nb** 92.91	42 **Mo** 95.94	43 **Tc** (98)	44 **Ru** 101.1	45 **Rh** 102.91	46 **Pd** 106.42	47 **Ag** 107.87	48 **Cd** 112.41	49 **In** 114.82	50 **Sn** 118.71	51 **Sb** 121.75	52 **Te** 127.60	53 **I** 126.91	54 **Xe** 131.29
55 **Cs** 132.91	56 **Ba** 137.33	57 ***La** 138.91	72 **Hf** 178.49	73 **Ta** 180.95	74 **W** 183.85	75 **Re** 186.21	76 **Os** 190.2	77 **Ir** 192.2	78 **Pt** 195.08	79 **Au** 196.97	80 **Hg** 200.59	81 **Tl** 204.38	82 **Pb** 207.2	83 **Bi** 208.98	84 **Po** (209)	85 **At** (210)	86 **Rn** (222)
87 **Fr** (223)	88 **Ra** 226.02	89 **†Ac** 227.03	104 **Rf** (261)	105 **Db** (262)	106 **Sg** (266)	107 **Bh** (264)	108 **Hs** (277)	109 **Mt** (268)	110 **Ds** (271)	111 **Rg** (272)							

	58 **Ce** 140.12	59 **Pr** 140.91	60 **Nd** 144.24	61 **Pm** (145)	62 **Sm** 150.4	63 **Eu** 151.97	64 **Gd** 157.25	65 **Tb** 158.93	66 **Dy** 162.50	67 **Ho** 164.93	68 **Er** 167.26	69 **Tm** 169.93	70 **Yb** 173.04	71 **Lu** 174.97
*Lanthanide Series														
†Actinide Series	90 **Th** 232.04	91 **Pa** 231.04	92 **U** 238.03	93 **X** (237)	94 **Pu** (244)	95 **Am** (243)	96 **Cm** (247)	97 **Bk** (247)	98 **Cf** (251)	99 **Es** (252)	100 **Fm** (257)	101 **Md** (258)	102 **No** (259)	103 **Lr** (262)

The periodic table gives you very basic but very important information about each element.

Bad Joke Alert

To prepare for the AP Chemistry Exam, familiarize yourself with the periodic table of elements by looking at it periodically.

1. This is the **symbol** for the element; carbon, in this case. On the test, the symbol for an element is used interchangeably with the name of the element.
2. This is the **atomic number** of the element. The atomic number is the same as the number of protons in the nucleus of an element; it is also the same as the number of electrons surrounding the nucleus of an element in its neutral state.
3. This is the **molar mass** of the element. It's also called the **atomic weight.**

The horizontal rows of the periodic table are called **periods.**

The vertical columns of the periodic table are called **groups.**

The vertical columns of the periodic table are called groups. Groups can be numbered in two ways. The old system used Roman numerals to indicate groups. The new system simply numbers the groups from 1 to 18. While it is not important to know the specific group numbers, it is important to know the names of some groups.

Group IA/1 – Alkali Metals
Group IIA/2 – Alkaline Earth Metals
Group B/3-12 – Transition Metals
Group VIIA/17 – Halogens
Group VIIIA/18 – Noble Gases

In addition, the two rows offset beneath the table are alternatively called the lanthanides and actinides, the rare earth elements, or the inner transition metals.

The identity of an atom is determined by the number of protons contained in its nucleus. The nucleus of an atom also contains neutrons. The mass number of an atom is the sum of its neutrons and protons. Atoms of an element with different numbers of neutrons are called isotopes; for instance, carbon-12, which contains 6 protons and 6 neutrons, and carbon-14, which contains 6 protons and 8 neutrons, are isotopes of carbon. The molar mass given on the periodic table is the average of the mass numbers of all known isotopes weighted by their percent abundance.

The mass of various isotopes of an element can be determined by a technique called mass spectrometry. A mass spec of selenium looks like this:

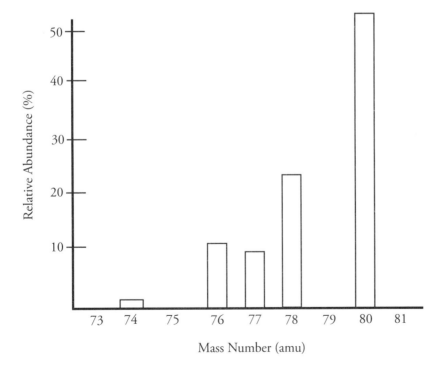

As you can see, the most abundant isotope of selenium has a mass of 80, but there are five other naturally occurring isotopes of selenium. The average atomic mass is the weighted average of all six isotopes of selenium shown on this spectra.

The molar mass of an element will give you a pretty good idea of the most common isotope of that element. For instance, the molar mass of carbon is 12.0 and about 99 percent of the carbon in existence is carbon-12.

MOLES

The mole (**Avogadro's number**) is the most important number in chemistry, serving as a bridge that connects all the different quantities that you'll come across in chemical calculations. The coefficients in chemical reactions tell you about the reactants and products in terms of moles, so most of the stoichiometry questions you'll see on the test will be exercises in converting between moles and grams, liters, molarities, and other units.

Moles and Molecules

The definition of Avogadro's number gives you the information you need to convert between moles and individual molecules and atoms.

$$1 \text{ mole} = 6.022 \times 10^{23} \text{ molecules}$$

$$\text{Moles} = \frac{\text{molecules}}{\left(6.022 \times 10^{23}\right)}$$

Moles and Grams

Moles and grams can be related using the atomic masses given in the periodic table. Atomic masses on the periodic table are given in terms of atomic mass units (amu), but an amu is the same as a gram per mole, so if 1 carbon atom has a mass of 12 amu, then 1 mole of carbon atoms has a mass of 12 grams.

You can use the relationship between amu and g/mol to convert between grams and moles by using the following equation:

$$\text{Moles} = \frac{\text{grams}}{\text{molar mass}}$$

Moles and Gases

We'll talk more about the ideal gas equation in Chapter 4, but for now, you should know that you can use it to calculate the number of moles of a gas if you know some of the gas's physical properties. All you need to remember at this point is that in the equation $PV = nRT$, n stands for moles of gas.

$$\text{Moles} = \frac{PV}{RT}$$

P = pressure (atm)
V = volume (L)
T = temperature (K)
R = the gas constant, 0.0821 L-atm/mol-K

The equation above gives the general rule for finding the number of moles of a gas. Many gas problems will take place at STP, or standard temperature and pressure, where $P = 1$ atmosphere and $T = 273$ K. At STP, the situation is much simpler and you can convert directly between the volume of a gas and the number of moles. That's because at STP, one mole of gas always occupies 22.4 liters.

$$\text{Moles} = \frac{\text{liters}}{\left(22.4\,\text{L/mol}\right)}$$

Moles and Solutions

We'll talk more about molarity and molality in Chapter 4, but for now you should realize that you can use the equations that define these common measures of concentration to find the number of moles of solute in a solution. Just rearrange the equations to isolate moles of solute.

Moles = (molarity)(liters of solution)
Moles = (molality)(kilograms of solvent)

Percent Composition

To solve many problems on the exam, you will need to use percent composition, or mass percents. Percent composition is the percent by mass of each element that makes up a compound. It is calculated by dividing the mass of each element or component in a compound by the total molar mass for the substance.

Calculate the percent composition of each element in calcium nitrate, $Ca(NO_3)_2$.

To do this, you need to first separate each element and count how many atoms are present. Subscripts outside of parentheses apply to all atoms inside of those parentheses.

Calcium: 1
Nitrogen: 2
Oxygen: 6

Then, multiply the number of atoms by the atomic mass of each element.

Ca: $1 \times 40.08 = 40.08$
N: $14.01 \times 2 = 28.02$
O: $16.00 \times 6 = 96.00$

Adding up the masses of the individual mass will give you the atomic mass of that compound. Divide each individual mass by the total atomic mass to get your percent composition.

$$40.08 + 28.02 + 96.00 = 164.10 \text{ g/mol}$$

Ca: $40.08/164.10 \times 100\% = 24.42\%$
N: $28.02/164.10 \times 100\% = 17.07\%$
O: $96.00/164.10 \times 100\% = 58.50\%$

You can check your work at the end by making sure your percents add up to 100% (taking rounding into consideration).

$24.42\% + 17.07\% + 58.50\% = 99.99\%$ Close enough!

Empirical and Molecular Formulas

You will also need to know how to determine the empirical and molecular formula of a compound given masses or mass percents of the components of that compound. Remember that the empirical formula represents the simplest ratio of one element to another in a compound (e.g., CH_2O), while the molecular formula represents the actual formula for the substance (e.g., $C_6H_{12}O_6$).

Electronic Configurations and the Periodic Table

The positively-charged nucleus is always pulling at the negatively-charged electrons around it, and the electrons have potential energy that increases with their distance from the nucleus. It works the same way that the gravitational potential energy of a brick on the third floor of a building is greater than the gravitational potential energy of a brick nearer to ground level.

The energy of electrons, however, is **quantized**. That's important. It means that electrons can exist only at specific energy levels, separated by specific intervals. It's kind of like if the brick in the building could be placed only on the first, second, or third floor of the building, but not in-between.

The Aufbau Principle

The **Aufbau principle** states that when building up the electron configuration of an atom, electrons are placed in orbitals, subshells, and shells in order of increasing energy.

The Pauli Exclusion Principle

The **Pauli exclusion principle** states that within an atom, no two electrons can have the same set of quantum numbers. So, each electron in any atom has its own distinct set of four quantum numbers.

Hund's Rule

Hund's rule says that when an electron is added to a subshell, it will always occupy an empty orbital if one is available. Electrons always occupy orbitals singly if possible and pair up only if no empty orbitals are available.

Watch how the $2p$ subshell fills as we go from boron to neon.

	1s	2s	2p
Boron	⇅	⇅	↑
Carbon	⇅	⇅	↑ ↑
Nitrogen	⇅	⇅	↑ ↑ ↑
Oxygen	⇅	⇅	⇅ ↑ ↑
Fluorine	⇅	⇅	⇅ ⇅ ↑
Neon	⇅	⇅	⇅ ⇅ ⇅

COULOMB'S LAW

The amount of energy that an electron has depends on its distance from the nucleus of the atom. This can be calculated using Coulomb's Law:

$$E = k + q - q/r$$

E = energy

k = Coulomb's constant

$+q$ = magnitude of the positive charge (nucleus)

$-q$ = magnitude of the negative charge (electron)

r = distance between the charges

While on the AP exam, you will not be required to mathematically calculate the amount of energy a given electron has, you should be able to qualitatively apply

Coulomb's Law. Essentially, the greater the charge of the nucleus, the more energy an electron will have (as all electrons have the same amount of charge). Also, the closer the electron is to the nucleus, the more energy it will have.

Quantum Theory

Max Planck figured out that electromagnetic energy is quantized. That is, for a given frequency of radiation (or light), all possible energies are multiples of a certain unit of energy, called a quantum (mathematically, that's $E = h\nu$). So, energy changes do not occur smoothly but rather in small but specific steps.

The Bohr Model

Neils Bohr took the quantum theory and used it to predict that electrons orbit the nucleus at specific, fixed radii, like planets orbiting the Sun.

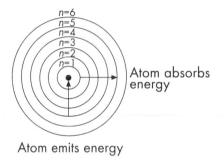

Each energy level is represent by a row on the periodic table. There are currently seven known energy levels, which correspond with $n = 1$ to $n = 7$. The closer an energy level is to an atom, the more energy electrons on that level have. Electrons further out not only have greater distance between themselves and the nucleus, but also are partially shielded from the nucleus by electrons on lower energy levels. While the Bohr model is not a perfect model of the atom, it serves as an excellent basis to understand atomic structure.

When atoms absorb energy in the form of **electromagnetic radiation**, electrons jump to higher energy levels. When electrons drop from higher to lower energy levels, atoms give off energy in the form of electromagnetic radiation.

The relationship between the change in energy level of an electron and the electromagnetic radiation absorbed or emitted is given below.

Energy and Electromagnetic Radiation

$$\Delta E = h\nu = \frac{hc}{\lambda}$$

ΔE = energy change
h = Planck's constant, 6.63×10^{-34} joule-sec
ν = frequency of the radiation
λ = wavelength of the radiation
c = the speed of light, 3.00×10^8 m/sec ($c = \lambda f$)

The energy level changes for the electrons of a particular atom are always the same, so atoms can be identified by their emission and absorption spectra.

Frequency and Wavelength
$$c = \lambda\nu$$

c = speed of light in a vacuum (2.998×10^8 ms^{-1})
λ = wavelength of the radiation
ν = frequency of the radiation

The frequency and wavelength of electromagnetic radiation are inversely proportional. Combined with the previous equation, we can see that higher frequencies and shorter wavelengths leads to more energy.

PHOTOELECTRON SPECTROSCOPY

If an atom is exposed to electromagnetic radiation of a high enough frequency, the energy of that radiation coming in may be enough to remove an electron from the atom. The amount of energy necessary to do that is called the **ionization energy** (also known as **binding energy**) for that electron. The closer an electron is to the nucleus, the higher its ionization energy will be. This energy can be measured in Joules (J) or electron volts (eV).

All energy of the incoming radiation must be conserved and any of that energy that does not go into breaking the electron free from the nucleus will be converted into **kinetic energy** (the energy of motion) for the ejected electron. So:

Incoming Energy (photon) = Ionization Energy + Kinetic energy (of the electron)

The faster an ejected electron is going, the less kinetic energy it has and the further it was from the nucleus originally. So, by examining the speed of the ejected electrons, we can determine how far they were from the nucleus of the atom in the first place. Usually, it takes electromagnetic radiation in either the visible or ultraviolet range to cause electron emission, while radiation in the infrared range is often used to study chemical bonds.

Spectra

If the amount of ionization energy for all electrons ejected from a nucleus is charted, you get what is called a **photoelectron spectra** (PES) that looks like this:

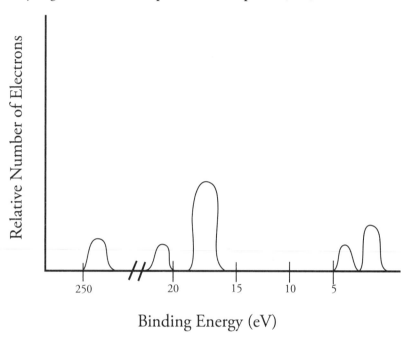

The *y*-axis describes the relative number of electrons that are ejected from a given energy level, and the *x*-axis shows the ionization energy of those electrons. Unlike most graphs, binding energy (ionization energy) decreases going from left to right in a PES. The spectra above is for sulfur.

Each section of peaks in the PES represents a different energy level. The number of peaks in a section shows us that not all electrons at *n* = 2+ are located the same distance from the nucleus. With each energy level, there are **subshells**, which describe the shape of the space the electron can be found in (remember, we are in three dimensional space here). The Bohr model is limited to two dimensions and does not represent the true positions of electrons due to that reason. Electrons do not orbit the nucleus as planets orbit the Sun. Instead, they are found moving about in a certain area of space (the subshell) a given distance away (the energy level) from the nucleus.

In energy levels 2 and higher, the first subshell is called the *s*-subshell and can hold a maximum of two electrons. The second subshell is called the *p*-subshell and can hold a maximum of six electrons. In the spectra above, we can see the peak for the *p*-subshell in energy level 2 is three times higher than that of the s-subshell. The relative height of the peaks helps determine the number of electrons in that subshell.

In the area for the third energy level, the *p*-subshell peak is only twice as tall as the *s*-subshell. This indicates there are only four electrons in the *p*-subshell of this particular atom.

Electron configuration

Studying the PES of elements allows scientists to understand more about the structure of the atom. In addition to the *s* and *p* subshells, two others exist: *d* (10 electrons max) and *f* (14 electrons max). The periodic table is designed so that each area is exactly the length of one particular subshell.

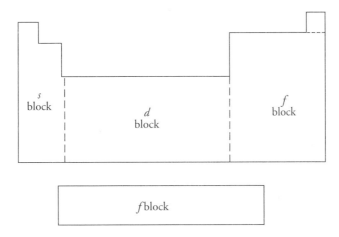

As you can see, the first two groups (plus helium) are in what is called the *s*-block. The groups on the right of the table are in the *p*-block, while transition metals make up the *d*-block. The inner transition metals below the table inhabit the *f*-block. The complete description of the energy level and subshell that each electron on an element inhabits is called its **electron configuration**.

Example 1: Determine the electron configuration for silicon.

Silicon has 16 electrons. The first two go into energy level 1 subshell *s*; this is represented by $1s^2$. The next two go into energy level 2 subshell *s*—$2s^2$. Six more fill energy level *p* ($2p^6$) then two more go into 3s ($3s^2$) and the final three enter into 3p ($3p^3$). So the final configuration is $1s^2 2s^2 2p^6 3s^2 3p^3$.

Using the periodic table as a reference can allow for the determination of any electron configuration. One thing to watch out for is that when entering the *d*-block (transition metals), the energy level drops by 1. The why behind that isn't important right now, but you should still be able to apply that rule.

Example 2: Determine the electron configuration for nickel.

$$1s^2 2s^2 2p^6 3s^2 3p^6 4s^2 3d^8$$

Electron configurations can also be written "shorthand" by replacing parts of them with the symbol for the noble gas at the end of the highest energy level which has been filled.

For example, the shorthand notation for Si is $[Ne]3s^23p^3$ and the shorthand for nickel would be $[Ar]4s^23d^8$.

Names and Theories

Dalton's Elements

In the early 1800s, John Dalton presented some basic ideas about atoms that we still use today. He was the first to say that there are many different kinds of atoms, which he called elements. He said that these elements combine to form compounds and that these compounds always contain the same ratios of elements. Water (H_2O), for instance, always has two hydrogen atoms for every oxygen atom. He also said that atoms are never created or destroyed in chemical reactions.

Development of the Periodic Table

In 1869, Dmitri Mendeleev and Lothar Meyer independently proposed arranging the elements into early versions of the periodic table, based on the trends of the known elements.

Thomson's Experiment

In the late 1800s, J. J. Thomson watched the deflection of charges in a cathode ray tube and put forth the idea that atoms are composed of positive and negative charges. The negative charges were called electrons, and Thomson guessed that they were sprinkled throughout the positively charged atom like chocolate chips sprinkled throughout a blob of cookie dough.

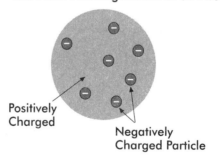

The Plum Pudding Model of an Atom

Millikan's Experiment

Robert Millikan was able to calculate the charge on an electron by examining the behavior of charged oil drops in an electric field.

Rutherford's Experiment

In the early 1900s, Ernest Rutherford fired alpha particles at gold foil and observed how they were scattered. This experiment led him to conclude that all of the positive charge in an atom was concentrated in the center and that an atom is mostly empty space. This led to the idea that an atom has a positively charged nucleus, which contains most of the atom's mass, and that the tiny, negatively charged electrons travel around this nucleus.

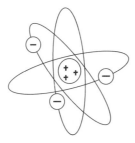

The Heisenberg Uncertainty Principle

Werner Heisenberg said that it is impossible to know both the position and momentum of an electron at a particular instant. In terms of atomic structure, this means that electron orbitals do not represent specific orbits like those of planets. Instead, an electron orbital is a probability function describing the possibility that an electron will be found in a region of space.

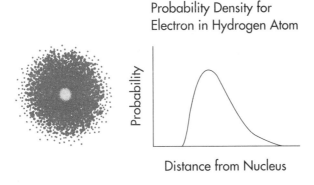

Probability Density for
Electron in Hydrogen Atom

Distance from Nucleus

PERIODIC TRENDS

You can make predictions about certain behavior patterns of an atom and its electrons based on the position of the atom in the periodic table. All the periodic trends can be understood in terms of three basic rules.

1. Electrons are attracted to the protons in the nucleus of an atom.
   a. The closer an electron is to the nucleus, the more strongly it is attracted.
   b. The more protons in a nucleus, the more strongly an electron is attracted.
2. Electrons are repelled by other electrons in an atom. So, if other electrons are between a valence electron and the nucleus, the valence electron will be less attracted to the nucleus. That's called shielding.
3. Completed shells (and to a lesser extent, completed subshells) are very stable. Atoms prefer to add or subtract valence electrons to create complete shells if possible.

The atoms in the left-hand side of the periodic table are called metals. Metals give up electrons when forming bonds. Most of the elements in the table are metals. The elements in the upper right-hand portion of the table are called nonmetals. Nonmetals generally gain electrons when forming bonds. The metallic character of the elements decreases as you move from left to right across the periodic table. The elements in the borderline between metal and nonmetal, such as silicon and arsenic, are called metalloids.

Atomic Radius

The atomic radius is the approximate distance from the nucleus of an atom to its valence electrons.

Moving from Left to Right Across a Period (Li to Ne, for Instance), Atomic Radius Decreases

Moving from left to right across a period, protons are added to the nucleus, so the valence electrons are more strongly attracted to the nucleus; this decreases the atomic radius. Electrons are also being added, but they are all in the same shell at about the same distance from the nucleus, so there is not much of a shielding effect.

Moving Down a Group (Li to Cs, for Instance), Atomic Radius Increases

Moving down a group, shells of electrons are added to the nucleus. Each shell shields the more distant shells from the nucleus and the valence electrons get farther away from the nucleus. Protons are also being added, but the shielding effect of the negatively charged electron shells cancels out the added positive charge.

Cations (Positively Charged Ions) Are Smaller than Atoms

Generally, when electrons are removed from an atom to form a cation, the outer shell is lost, making the cation smaller than the atom. Also, when electrons are removed, electron-electron repulsions are reduced, allowing all of the remaining valence electrons to move closer to the nucleus.

Anions (Negatively Charged Ions) Are Larger than Atoms

When an electron is added to an atom, forming an anion, electron-electron repulsions increase, causing the valence electrons to move farther apart, which increases the radius.

Ionization Energy

Electrons are attracted to the nucleus of an atom, so it takes energy to remove an electron. The energy required to remove an electron from an atom is called the first ionization energy. Once an electron has been removed, the atom becomes a positively charged ion. The energy required to remove the next electron from the ion is called the second ionization energy, and so on.

Moving from Left to Right Across a Period, Ionization Energy Increases

Moving from left to right across a period, protons are added to the nucleus, which increases its positive charge. For this reason, the negatively charged valence electrons are more strongly attracted to the nucleus, which increases the energy required to remove them.

Moving Down a Group, Ionization Energy Decreases

Moving down a group, shells of electrons are added to the nucleus. Each inner shell shields the more distant shells from the nucleus, reducing the pull of the nucleus on the valence electrons and making them easier to remove. Protons are also being added, but the shielding effect of the negatively charged electron shells cancels out the added positive charge.

The Second Ionization Energy Is Greater than the First Ionization Energy

When an electron has been removed from an atom, electron-electron repulsion decreases and the remaining valence electrons move closer to the nucleus. This increases the attractive force between the electrons and the nucleus, increasing the ionization energy.

As Electrons Are Removed, Ionization Energy Increases Gradually Until a Shell Is Empty, Then Makes a Big Jump

- For each element, when the valence shell is empty, the next electron must come from a shell that is much closer to the nucleus, making the ionization energy for that electron much larger than for the previous ones.
- For Na, the second ionization energy is much larger than the first.
- For Mg, the first and second ionization energies are comparable, but the third is much larger than the second.
- For Al, the first three ionization energies are comparable, but the fourth is much larger than the third.

Electronegativity

Electronegativity refers to how strongly the nucleus of an atom attracts the electrons of other atoms in a bond. Electronegativity is affected by two factors. The smaller an atom is, the more effectively its nuclear charge will be felt past its outermost energy level and the higher its electronegativity will be. Second, the closer an element is to having a full energy level, the more likely it is to attract the necessary electrons to complete that level. In general:

- Moving from left to right across a period, electronegativity increases.
- Moving down a group, electronegativity decreases.

The various periodic trends, which don't include the noble gases, are summarized in the diagram below.

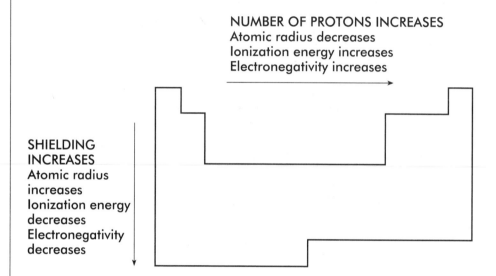

NUMBER OF PROTONS INCREASES
Atomic radius decreases
Ionization energy increases
Electronegativity increases

SHIELDING INCREASES
Atomic radius increases
Ionization energy decreases
Electronegativity decreases

CHAPTER 3 QUESTIONS

Multiple-Choice Questions

Use the PES spectra below to answer questions 1-4.

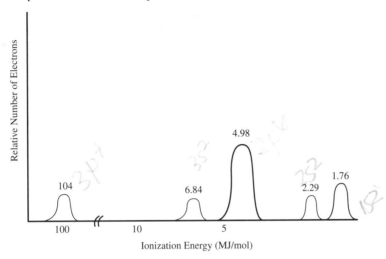

1. What element does this spectra represent?

    (A) Boron
    (B) Nitrogen
    (C) Aluminum
    (D) Phosphorus

2. Which peak represents the 2s subshell?

    (A) The peak at 104 MJ/mol
    (B) The peak at 6.84 MJ/mol
    (C) The peak at 2.29 MJ/mol
    (D) The peak at 1.76 MJ/mol

3. An electron from which peak would have the most Coulombic potential energy?

    (A) The peak at 104 MJ/mol
    (B) The peak at 6.84 MJ/mol
    (C) The peak at 4.98 MJ/mol
    (D) The peak at 1.76 MJ/mol

4. How many valence electrons does this atom have?

    (A) 2
    (B) 3
    (C) 4
    (D) 5

5. Why does an ion of phosphorus, P^{3-}, have a larger radius than a neutral atom of phosphorus?

   (A) There is a greater Coulombic attraction between the nucleus and the electrons in P^{3-}.
   (B) The core electrons in P^{3-} exert a weaker shielding force than those of a neutral atom.
   (C) The nuclear charge is weaker in P^{3-} than it is in P.
   (D) The electrons in P^{3-} have a greater Coulombic repulsion than those in the neutral atom.

6. A binary compound containing only copper and iodine is found to be 16.5% copper by mass. What is the empirical formula of this compound?

   (A) CuI
   (B) CuI_2
   (C) CuI_3
   (D) CuI_4

7. Which neutral atom of the following elements would have the most unpaired electrons?

   (A) Titanium
   (B) Manganese
   (C) Nickel
   (D) Zinc

8. Which element will have a higher electronegativity value: chlorine or bromine? Why?

   (A) Chlorine, because it has less Coulombic repulsion among its electrons
   (B) Bromine, because it has more protons
   (C) Chlorine, because it is smaller
   (D) Bromine, because it is larger

9. Which of the following elements has its highest energy subshell completely full?

   (A) Sodium
   (B) Aluminum
   (C) Chlorine
   (D) Zinc

10. Which of the following isoelectric species has the smallest radius?

    (A) S^{2-}
    (B) Cl^-
    (C) Ar
    (D) K^+

11. What is the most likely electron configuration for a sodium ion in its ground state?

    (A) $1s^2\, 2s^2\, 2p^5$
    (B) $1s^2\, 2s^2\, 2p^6$
    (C) $1s^2\, 2s^2\, 2p^6\, 3s^1$
    (D) $1s^2\, 2s^2\, 2p^5\, 3s^2$

12. Which of the following statements is true regarding sodium and chlorine?

(A) Sodium has greater electronegativity and a larger first ionization energy.
(B) Sodium has a larger first ionization energy and a larger atomic radius.
(C) Chlorine has a larger atomic radius and a greater electronegativity.
(D) Chlorine has greater electronegativity and a larger first ionization energy.

13. An atom of silicon in its ground state is subjected to a frequency of light that is high enough to cause electron ejection. An electron from which subshell of silicon would have the highest kinetic energy after ejection?

(A) $1s$
(B) $2p$
(C) $3p$
(D) $4s$

14. The wavelength range for infrared radiation is 10^{-5} m, while that of ultraviolet radiation is 10^{-8} m. Which type of radiation has more energy, and why?

(A) Ultraviolet, because it has a higher frequency
(B) Ultraviolet, because it has a longer wavelength
(C) Infrared, because it has a lower frequency
(D) Infrared, because it has a shorter wavelength

15. Which of the following nuclei has 3 more neutrons than protons? (Remember: The number before the symbol indicates atomic mass.)

(A) ^{11}B
(B) ^{37}Cl
(C) ^{24}Mg
(D) ^{70}Ga

16. Which of the following ions has the smallest ionic radius?

(A) O^{2-}
(B) F^{-}
(C) Mg^{2+}
(D) Al^{3+}

17. Which of the following is true of the halogens when comparing them to other elements in the same period?

(A) They have larger atomic radii than other elements within their period.
(B) They have a lower ionization energy than elements within their period.
(C) They have less peaks on a PES than other elements within their period.
(D) Their electronegativity is higher than other elements within their period.

18. In general, do metals or nonmetals from the same period have higher ionization energies? Why?

(A) Metals, because they usually have more protons than nonmetals
(B) Nonmetals, because they are larger than metals and harder to ionize
(C) Metals, because there is less electron shielding than there is in nonmetals
(D) Nonmetals, because they are closer to having filled a complete energy level

19. The ionization energies for an element are listed in the table below.

First   Second   Third   Fourth   Fifth
8 eV   15 eV   80 eV   109 eV   141 eV

Based on the ionization energy table, the element is most likely to be

(A) sodium.
(B) magnesium.
(C) aluminum.
(D) silicon.

20. A researcher listed the first five ionization energies for a silicon atom in order from first to fifth. Which of the following lists corresponds to the ionization energies for silicon?

(A) 780 kJ, 13,675 kJ, 14,110 kJ, 15,650 kJ, 16,100 kJ
(B) 780 kJ, 1,575 kJ, 14,110 kJ, 15,650 kJ, 16,100 kJ
(C) 780 kJ, 1,575 kJ, 3,220 kJ, 15,650 kJ, 16,100 kJ
(D) 780 kJ, 1,575 kJ, 3,220 kJ, 4,350 kJ, 16,100 kJ

Constructed Response Questions

1. Explain each of the following in terms of atomic and molecular structures and/or forces.

   (a) The first ionization energy for magnesium is greater than the first ionization energy for calcium.
   (b) The first and second ionization energies for calcium are comparable, but the third ionization energy is much greater.
   (c) There are three peaks of equal height in the PES of carbon, but on the PES of oxygen the last peak has a height twice as high as all the others.
   (d) The first ionization energy for aluminum is lower than the first ionization energy for magnesium.

2. Use your knowledge of the periodic table of the elements to answer the following questions.

   (a) Explain the trend in electronegativity from P to S to Cl.
   (b) Explain the trend in electronegativity from Cl to Br to I.
   (c) Explain the trend in atomic radius from Li to Na to K.
   (d) Explain the trend in atomic radius from Al to Mg to Na.

3. The table below gives data on four different elements, in no particular order:

Carbon, Oxygen, Phosphorus, and Chlorine

	Atomic radius (pm)	First Ionization Energy (kJ mol-1)
Element 1	170	1086.5
Element 2	180	1011.8
Element 3	175	1251.2
Element 4	152	1313.9

(a) Which element is number 3? Justify your answer using both properties.
(b) What is the outermost energy level that has electrons in element 2? How many valence electrons does element 2 have?
(c) Which element would you expect to have the highest electronegativity? Why?
(d) How many peaks would the PES for element 4 have and what would the relative heights of those peaks be to each other?

CHAPTER 3 ANSWERS AND EXPLANATIONS

Multiple-Choice Questions

1. **D** This element has five peaks, meaning a total of five subshells. The final peak, which would be located in the 3*p* subshell, is slightly higher than the 3*s* peak to the left of it. A full 3*s* peak has two electrons, therefore there must be at least three electrons in the 3*p* subshell.

2. **B** The peaks, in order, represent 1*s*, 2*s*, 2*p*, 3*s*, and 3*p*.

3. **A** The *x*-axis represents ionization energy in MJ/mol. Ionization energy has the same magnitude as Coulombic potential energy. Separately, the first peak represents the 1*s* subshell, which is closest to the nucleus. Using Coulomb's law, the smaller the distance is between the nucleus and the electron, the greater the potential energy is.

4. **D** Valence electrons are defined as those in the highest occupied energy level. The two peaks in energy level three which have electrons here are 3*s*, which has two electrons, and 3*p*, which has three electrons. The total number of valence electrons is thus: 2 + 3 = 5.

5. **D** The ion has three more electrons than the neutral atom, meaning the overall repulsion will be greater. The electrons will "push" each other away more effectively, creating a bigger radius.

6. **B** Copper has a mass of 63; iodine has a mass of 126. One atom of each creates a compound that is approximately 33% copper by mass. One atom of copper and two of iodine creates a compound that is half that, or 16.5%, by mass.

7. **B** In all of these transition metals, the 3*d* subshell is the last one to receive electrons (Aufbau's Principle). A 3*d* subshell has 5 orbitals which can hold two electrons each (Pauli's Exclusion Principle). The electrons will enter one into each orbital before any pair up (Hund's Rule), meaning manganese will have all five of its 3*d* electrons unpaired.

8. **C** The smaller an atom is, the more the "pull" of the nucleus can be felt and the easier it will be for that element to attract more electrons. (B) is a tempting answer, however, the additional protons in bromine are much further away from the outside of the atom, as it has one full energy level greater than chlorine.

9. **D** Zinc has 10 electrons in the $3d$ subshell, filling it completely.

10. **D** Potassium has the highest number of protons out of all the options, therefore it exerts a higher nuclear charge on the electrons, pulling them in closer and creating a smaller atomic radius.

11. **B** Neutral sodium in its ground state has the electron configuration shown in choice (C). Sodium forms a bond by giving up its one valence electron and becoming a positively charged ion with the same electron configuration as neon.

12. **D** As we move from left to right across the periodic table within a single period (from sodium to chlorine), we add protons to the nuclei, which progressively increases the pull of each nucleus on its electrons. So chlorine will have a higher first ionization energy, greater electronegativity, and a smaller atomic radius.

13. **C** The further an electron is from the nucleus, the less binding energy it has and the easier it is to eject. The electrons in $3p$ would have the lowest binding energy and thus the highest velocity (and kinetic energy). A $4s$ electron would be faster; however, silicon has no electrons in the $4s$ subshell while in its ground state.

14. **A** Via $c = v\lambda$, we can see that there is an inverse relationship between wavelength and frequency (as one goes up, the other goes down). So, ultraviolet radiation has a higher frequency, and via $E = hv$, also has more energy.

15. **B** The atomic mass is the sum of the neutrons and protons in any atom's nucleus. Since atomic number, which indicates the number of protons, is unique to each element we can subtract this from the weight to find the number of neutrons. The atomic number of B is 5; hence, the ^{11}B has 6 neutrons, only 1 in excess. The atomic number of Cl is 17, so ^{37}Cl has 20 neutrons, 3 in excess. The atomic number of Mg is 12, so ^{24}Mg has the same number of neutrons as protons. The element Ga has an atomic number of 31, meaning there are 39 neutrons in the given nucleus and the atomic number for F is 9, meaning 10 neutrons in ^{19}F.

16. **D** All of the ions listed have the same electron configuration as neutral neon. Al^{3+} has the most protons, so its electrons will experience greater attractive force from the nucleus, resulting in the smallest ions.

17. **D** Electronegativity describes how easy it is for an element to attract additional electrons. Because halogens are smaller than other elements in their period (save for the noble gases, which have full energy levels and are not likely to gain additional electrons), they tend to have the highest electronegativity values within their period.

18. **D** Nonmetals appear on the right side of the periodic table, and so are closer to having a full energy level. Nonmetals are much more likely to gain electrons to fill their levels, and it is difficult to remove their electrons for that reason.

19. **B** The ionization energy will show a large jump when enough electrons have been removed to leave a stable shell. In this case, the jump occurs between the second and third electrons removed, so the element is stable after two electrons are removed. Magnesium (Mg) is the only element on the list with exactly two valence electrons.

20. **D** The ionization energy will show a large jump when enough electrons have been removed to leave a stable shell. Silicon has four valence electrons, so we would expect to see a large jump in ionization energy after the fourth electron has been removed. That's choice (D).

Constructed Response Questions

1. (a) Ionization energy is the energy required to remove an electron from an atom. The outermost electron in Ca is at the $4s$ energy level. The outermost electron in Mg is at the $3s$ level. The outermost electron in Ca is at a higher energy level and is more shielded from the nucleus, making it easier to remove.

    (b) Calcium has two electrons in its outer shell. The second ionization energy will be larger than the first but still comparable because both electrons are being removed from the same energy level. The third electron is much more difficult to remove because it is being removed from a lower energy level, so it will have a much higher ionization energy than the other two.

    (c) The height of the peaks on a PES represent the relative number of electrons in each subshell. In carbon, all three subshells hold two electron ($1s^2 2s^2 2p^2$), and thus all peaks are the same height. In oxygen, the $2p$ subshell has four electrons, meaning its peak will be twice as high as the other two.

(d) The valence electron to be removed from magnesium is located in the completed 3s subshell, while the electron to be removed from aluminum is the lone electron in the 3p subshell. It is easier to remove the electron from the higher-energy 3p subshell than from the lower energy (completed) 3s subshell, so the first ionization energy is lower for aluminum.

2. (a) Electronegativity is the pull of the nucleus of one atom on the electrons of other atoms; it increases from P to S to Cl because nuclear charge increases. This is because as you from move left to right across the periodic table, atomic radii decrease in size. Increasing nuclear charge means that Cl has the most positively charged nucleus of the three and will exert the greatest pull on the electrons of other atoms.

(b) Electronegativity is the pull of the nucleus of one atom on the electrons of other atoms; it decreases from Cl to Br to I because electron shells are added. The added electron shells shield the nucleus, causing it to have less of an effect on the electrons of other atoms. Therefore, iodine will exert the least pull on the electrons of other atoms.

(c) Atomic radius increases from Li to Na to K because electrons are being added in higher energy levels, which are farther away from the nucleus; therefore, the K atom is the largest of the three.

(d) Atomic radius increases from Al to Mg to Na because protons are being removed from the nucleus while the energy levels of the valence electrons remains unchanged. If there are fewer positive charges in the nucleus, the electrons of Na will be less attracted to the nucleus and will remain farther away.

3. (a) Element 3 is chlorine. Chlorine and phosphorus would have the largest atomic radii as they both have three energy levels with electrons present. However, chlorine would be smaller than phosphorus because it has more protons (a higher effective nuclear charge). Additionally, chlorine would have a higher ionization energy than phosphorus due to its smaller size and greater number of protons.

(b) Element 2 is phosphorus, and therefore the outermost energy level would be $n = 3$. Phosphorus has two electrons in 3s and three electrons in 3p for a total of five valence electrons.

(c) Electronegativity increases as atomic radius decreases, so it is expect that element 4 (oxygen) would have the highest electronegativity value. Alternatively, electronegativity increases as an energy level comes close to being full, so it is possible that element 3 (chlorine) may have the highest electronegativity as it is only one electron away from filling its outermost energy level. (Either answer is acceptable with the proper justification)

(d) Element 4 is oxygen, so it would be expect to have three peaks in a PES spectra, one for each subshell. The first two peaks would be the same height because there are two electrons each in the $1s$ and $2s$ subshells. The final peak would be twice the height of the others are there are four electrons in oxygen's $2p$ subshell.

Chapter 4
Big Idea #2:
Bonding and Phases

Chemical and physical properties of materials can be explained by the structure and the arrangement of atoms, ions, or molecules and the forces between them.

BONDS OVERVIEW

Atoms join to form molecules because atoms like to have a full outer shell of electrons. This usually means having eight electrons in the outer shell. So atoms with too many or too few electrons in their valence shells will find one another and pass the electrons around until all the atoms in the molecule have stable outer shells. Sometimes an atom will give up electrons completely to another atom, forming an ionic bond. Sometimes atoms share electrons, forming covalent bonds.

Ionic Bonds

An ionic solid is held together by the electrostatic attractions between ions that are next to one another in a lattice structure. They often occur between metals and nonmetals. In an ionic bond, electrons are not shared. Instead, one atom gives up electrons and becomes a positively charged ion while the other atom accepts electrons and becomes a negatively charged ion.

The two ions in an ionic bond are held together by electrostatic forces. In the diagram below, a sodium atom has given up its single valence electron to a chlorine atom, which has seven valence electrons and uses the electron to complete its outer shell (with eight). The two atoms are then held together by the positive and negative charges on the ions.

$$\left[Na \right]^+ \left[:\ddot{C}\ddot{l}: \right]^-$$

The electrostatic attractions that hold together the ions in the NaCl lattice are very strong and any substance held together by ionc bonds will usually be a solid at room temperature and have very high melting and boiling points.

Two factors affect the melting points of ionic substances. The primary factor is the charge on the ions. According to Coulomb's law, a greater charge leads to a greater bond energy (often called lattice energy in ionic bonds), so a compound composed of ions with charges of +2 and –2 (such as MgO) will have a higher melting point than a compound composed of ions with charges of +1 and –1 (such as NaCl). If both compounds are made up of ions with equal charges, then the size of the ions must be considered. Smaller ions will have greater Coulombic attraction (remember, size is inversely proportional to bond energy), so a substance like LiF would have a greater melting point than KBr.

In an ionic solid, each electron is localized around a particular atom, so electrons do not move around the lattice; this makes ionic solids poor conductors of electricity. Ionic liquids, however, do conduct electricity because the ions themselves are free to move about in the liquid phase, although the electrons are still localized around particular atoms. Salts are held together by ionic bonds.

Metallic Bonds

When examining metals, the sea of electrons model can be used. The positively-charged core of a metal, consisting of its nucleus and core electrons, are generally stationary, while the valence electrons on each atom do not belong to a specific atom and are very mobile. These mobile electrons explain why metals are such good conductors of electricity. The delocalized structure of a metal also explains why metals are both malleable and ductile, as deforming the metal does not change the environment immediately surrounding the metal cores.

Metals can also bond with each other to form alloys. In an **interstitial alloy**, metal atoms with two vastly different radii combine. Steel is one example; the much smaller carbon atoms occupy the interstices of the iron atoms. A **substitutional alloy** forms between atoms of similar radii. Brass is a good example, atoms of zinc are substituted for some copper atoms to create the alloy.

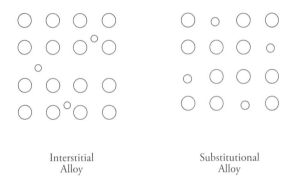

Interstitial
Alloy

Substitutional
Alloy

Covalent Bonds

In a covalent bond, two atoms share electrons. Each atom counts the shared electrons as part of its valence shell. In this way, both atoms achieve complete outer shells.

In the diagram below, two fluorine atoms, each of which has seven valence electrons and needs one electron to complete its valence shell, form a covalent bond. Each atom donates an electron to the bond, which is considered to be part of the valence shell of both atoms.

$$:\ddot{\text{F}}\cdot \;+\; \cdot\ddot{\text{F}}: \;\Rightarrow\; :\ddot{\text{F}}:\ddot{\text{F}}:$$

The number of covalent bonds an atom can form is the same as the number of electrons in its valence shell.

Single bonds have one sigma (σ) bond and a bond order of one. The single bond has the longest bond length and the least bond energy.

The first covalent bond formed between two atoms is called a sigma (σ) bond. All single bonds are sigma bonds. If additional bonds between the two atoms are formed, they are called pi (π) bonds. The second bond in a double bond is a pi bond and the second and third bonds in a triple bond are also pi bonds. Double and triple bonds are stronger and shorter than single bonds, but they are not twice or triple the strength.

Summary of Multiple Bonds			
Bond type:	Single	Double	Triple
Bond designation:	One sigma (s)	One sigma (σ) and one pi (π)	One sigma (σ) and two pi (π)
Bond order:	One	Two	Three
Bond length:	Longest	Intermediate	Shortest
Bond energy:	Least	Intermediate	Greatest

Polarity

In the F_2 molecule shown above, the two fluorine atoms share the electrons equally, but that's not usually the case in molecules. Usually, one of the atoms (the more electronegative one) will exert a stronger pull on the electrons in the bond—not enough to make the bond ionic, but enough to keep the electrons on one side of the molecule more than on the other side. This gives the molecule a dipole. That is, the side of the molecule where the electrons spend more time will be negative and the side of the molecule where the electrons spend less time will be positive.

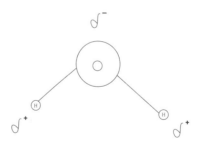

In the water molecule above, oxygen has a higher electronegativity than hydrogen and thus will have the electrons closer to it more often. This gives the oxygen a negative dipole and each hydrogen a positive dipole.

Dipole Moment

The polarity of a molecule is measured by the dipole moment. The larger the dipole moment, the more polar the molecule. The greater the charge at the ends of the dipole and the greater the distance between the charges, the greater the value of the dipole moment.

Intermolecular Forces

Sometimes the bonds that hold the atoms or ions in liquids and solids together are the same strong bonds that hold the atoms or ions together in molecules. Intermolecular forces exist only with covalently bonded molecules.

Network (Covalent) Bonds

In a network solid, atoms are held together in a lattice of covalent bonds. You can visualize a network solid as one big molecule. Network solids are very hard and have very high melting and boiling points.

The electrons in a network solid are localized in covalent bonds between particular atoms, so they are not free to move about the lattice. This makes network solids poor conductors of electricity.

The most commonly seen network solids are compounds of carbon (diamond) and silicon (SiO_2—quartz). The hardness of diamond is due to the tetrahedral network structure formed by carbon atoms whose electrons are configured in sp^3 hybridization. The three-dimensional complexity of the tetrahedral network means that there are no natural seams along which a diamond can be broken.

Silicon also serves as a semiconductor when it is doped with other elements. **Doping** is a process in which an impurity is added to an existing lattice. In a normal silicon lattice, each individual silicon atom is bonded to four other silicon atoms. When some silicon atoms are replaced with elements that have only three valence electrons (such as boron or aluminum), that means that the neighboring silicon atoms will lack one bond apiece.

This missing bond (or "hole") creates a positive charge in the lattice, and the hole attracts other electrons to it, increasing conductivity. Those electrons leave behind holes when they move, creating a chain reaction in which the conductivity of the silicon increases. This type of doping is called p-doping for the positively charged holes.

If an element with five valence electrons (such as phosphorus or arsenic) is used to add impurities to a silicon lattice, there is an extra valence electron that is free to move around the lattice, causing an overall negative charge which increases the conductivity of the silicon. This type of doping is called n-doping due to those free moving negatively charged electrons.

Hydrogen Bonds

Hydrogen bonds are similar to dipole–dipole attractions. In a hydrogen bond, the positively charged hydrogen end of a molecule is attracted to the negatively charged end of another molecule containing an extremely electronegative element (fluorine, oxygen, or nitrogen—F, O, N).

Hydrogen bonds are much stronger than dipole–dipole forces because when a hydrogen atom gives up its lone electron to a bond, its positively charged nucleus is left virtually unshielded. Substances that have hydrogen bonds, like water and ammonia, have higher melting and boiling points than substances that are held together by dipole–dipole forces.

Water is less dense as a solid than as a liquid because its hydrogen bonds force the molecules in ice to form a crystal structure, which keeps them farther apart than they are in the liquid form.

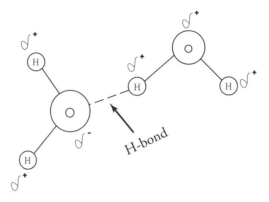

Dipole–Dipole Forces

Dipole–dipole forces occur between neutral, polar molecules: The positive end of one polar molecule is attracted to the negative end of another polar molecule.

Molecules with greater polarity will have greater dipole–dipole attraction, so molecules with larger dipole moments tend to have higher melting and boiling points. Dipole–dipole attractions are relatively weak, however, and these substances melt and boil at very low temperatures. Most substances held together by dipole–dipole attraction are gases or liquids at room temperature.

London Dispersion Forces

London dispersion forces occur between neutral, nonpolar molecules. These very weak attractions occur because of the random motions of electrons on atoms within molecules. At a given moment, a nonpolar molecule might have more electrons on one side than on the other, giving it an instantaneous polarity. For that fleeting instant, the molecule will act as a very weak dipole.

Since London dispersion forces depend on the random motions of electrons, molecules with more electrons will experience greater London dispersion forces. So among substances that experience only London dispersion forces, the one with more electrons will generally have higher melting and boiling points. London dispersion forces are even weaker than dipole–dipole forces, so substances that experience only London dispersion forces melt and boil at extremely low temperatures and tend to be gases at room temperature.

Melting Points Summary:

The melting (or boiling) point of a covalent substance is entirely based on the strength of its intermolecular forces, but the melting point of an ionic substance is based on its lattice energy. When a covalent substance undergoes a phase change, the intramolecular bonds within the molecules themselves are not breaking—just the intermolecular forces that hold the molecules together. When an ionic substance undergoes a phase change, the bonds that hold the lattice together are actually breaking. A summary of the different types of bonds and their relative melting boiling points from highest to lowest:

1. Network Covalent Bonds (usually SiO_2, C(diamond), or C(graphite))
2. Ionic Bonds (based on Coulombic attraction)
   a. Greater Ion Charge
   b. Smaller Atomic Size
3. Covalent Bonds (based on molecule polarity)
   a. Hydrogen Bonds
   b. Non-Hydrogen Bond Dipoles
   c. London Dispersion Forces (temporary dipoles)
      i. Large Molecules are more polarizable because they have more electrons

LEWIS DOT STRUCTURES

Drawing Lewis Dot Structures

At some point on the test, you'll be asked to draw the Lewis structure for a molecule or polyatomic ion. Here's how to do it.

1. Count the valence electrons in the molecule or polyatomic ion; refer to page 24 for a periodic table.
2. If a polyatomic ion has a negative charge, add electrons equal to the charge of the total in (1). If a polyatomic ion has a positive charge, subtract electrons equal to the charge of the electrons from the total in (1).
3. Draw the skeletal structure of the molecule and place two electrons (or a single bond) between each pair of bonded atoms. If the molecule contains three or more atoms, the least electronegative atom will usually occupy the central position.
4. Add electrons to the surrounding atoms until each has a complete outer shell.
5. Add the remaining electrons to the central atom.

6. Look at the central atom.
   (a) If the central atom has fewer than eight electrons, remove an electron pair from an outer atom and add another bond between that outer atom and the central atom. Do this until the central atom has a complete octet.
   (b) If the central atom has a complete octet, you are finished.
   (c) If the central atom has more than eight electrons, that's okay too.

Let's find the Lewis structure for the CO_3^{2-} ion.

1. Carbon has 4 valence electrons; oxygen has 6.
   $4 + 6 + 6 + 6 = 22$
2. The ion has a charge of –2, so add 2 electrons.
   $22 + 2 = 24$
3. Carbon is the central atom.

4. Add electrons to the oxygen atoms.

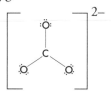

5. We've added all 24 electrons, so there's nothing left to put on the carbon atom.
6. (a) We need to give carbon a complete octet, so we take an electron pair away from one of the oxygens and make a double bond instead. Place a bracket around the model and add a charge of negative two.

Resonance Forms

When we put a double bond into the CO_3^{2-} ion, we place it on any one of the oxygen atoms, as shown below.

All three resonance forms are considered to exist simultaneously, and the strength and lengths of all three bonds are the same: somewhere between the strength and length of a single bond and a double bond.

To determine the relative length and strength of a bond in a resonance structure, a bond order calculation can be used. A single bond has a bond order of 1, and a double bond has an order of 2. When resonance occurs, pick one of the bonds in the resonance structure and add up the total bond order across the resonance forms, then divide that sum by the number of resonance forms.

For example, in the carbonate ion above, the top C–O bond would have a bond order of $(1 + 2 + 1)/3$, or 1.33. Bond order can be used to compare the length and strength of resonance bonds with pure bonds as well as other resonance bonds.

Incomplete Octets

Some atoms are stable with less than eight electrons in their outer shell. Hydrogen only requires two electrons, as does helium (although helium never forms bonds). Boron is considered to be stable with six electrons, as in the BF_3 diagram below. All other atoms involved in covalent bonding require a minimum of eight electrons to be considered stable.

Boron can be stable with only six valence electrons, as in BF_3.

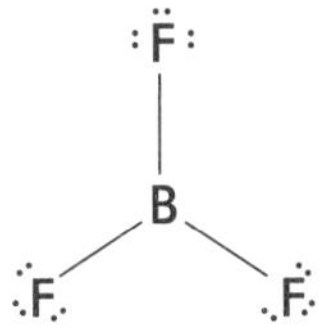

Expanded Octets

In molecules that have d subshells available, the central atom can have more than eight valence electrons, but never more than twelve. This means any atom of an element from $n = 3$ or greater can have expanded octets, but NEVER elements in $n = 2$ (C, N, O, etc.). Expanded octets also explains why some noble gases can actually form bonds; the extra electrons go into the empty d-orbital.

Here are some examples.

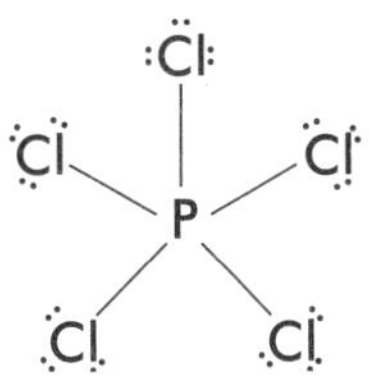

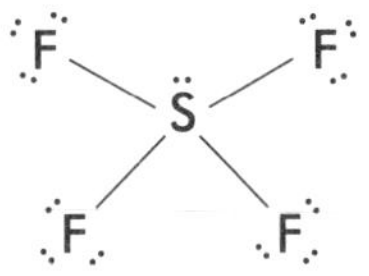

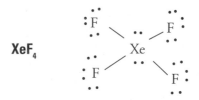

XeF_4

Formal Charge

Sometimes, there is more than one valid Lewis structure for a molecule. Take CO_2; it has two valid structures as shown below. To determine the more likely structure, formal charge is used. To calculate the formal charge on atoms in a molecule, take the number of valence electrons for that atom and subtract the number of assigned electrons in the Lewis structure. When counting assigned electrons, lone pairs count as 2 and bonds count as one.

O = C = O				:Ö — C ≡ O:		
6	4	6	valence e⁻	6	4	6
⁻6	⁻4	⁻6	assigned e⁻	⁻7	⁻4	⁻5
0	0	0	formal charge	−1	0	+1

The formal charge for a neutral molecule should be zero, which it is on both diagrams. Additionally, the less atoms that have an actual formal charge, the more likely the structure will be- so the left structure is the more likely one for CO_2. For polyatomic ions, the sum of the formal charges on each atom should equal the overall charge on the ion."

Molecular Geometry

Electrons repel one another, so when atoms come together to form a molecule, the molecule will assume the shape that keeps its different electron pairs as far apart as possible. When we predict the geometries of molecules using this idea, we are using the **valence shell electron-pair repulsion (VSEPR) model.**

In a molecule with more than two atoms, the shape of the molecule is determined by the number of electron pairs on the central atom. The central atom forms **hybrid orbitals,** each of which has a standard shape. Variations on the standard shape occur depending on the number of bonding pairs and lone pairs of electrons on the central atom.

Here are some things you should remember when dealing with the VSEPR model.

- Double and triple bonds are treated in the same way as single bonds in terms of predicting overall geometry for a molecule; however, multiple bonds have slightly more repulsive strength and will therefore occupy a little more space than single bonds.
- Lone electron pairs have a little more repulsive strength than bonding pairs, so lone pairs will occupy a little more space than bonding pairs.

The tables on the following pages show the different hybridizations and geometries that you might see on the test.

If the central atom has **2** electron pairs, then it has *sp* hybridization and its basic shape is **linear.**

Number of lone pairs	Geometry	Examples
0	B—A—B	$BeCl_2$
	linear	CO_2

If the central atom has **3** electron pairs, then it has *sp²* hybridization and its basic shape is **trigonal planar**; its bond angles are about 120°.

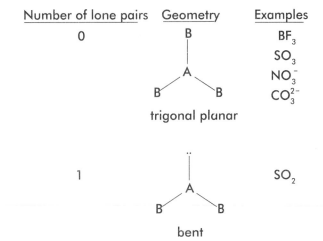

Number of lone pairs	Geometry	Examples
0	trigonal planar	BF_3 SO_3 NO_3^- CO_3^{2-}
1	bent	SO_2

If the central atom has 4 electron pairs, then it has sp^3 hybridization and its basic shape is **tetrahedral**; its bond angles are about 109.5°.

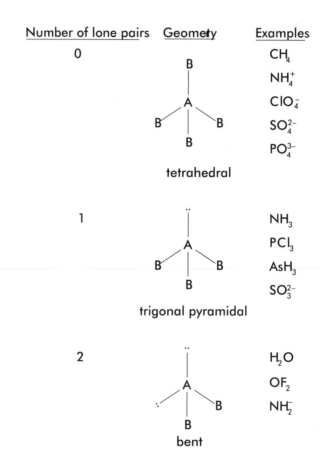

Number of lone pairs	Geometry	Examples
0	tetrahedral	CH_4 NH_4^+ ClO_4^- SO_4^{2-} PO_4^{3-}
1	trigonal pyramidal	NH_3 PCl_3 AsH_3 SO_3^{2-}
2	bent	H_2O OF_2 NH_2^-

If the central atom has 5 electron pairs, then it has dsp^3 hybridization and its basic shape is **trigonal bipyramidal.**

Number of lone pairs	Geometry	Examples
0		PCl_5
		PF_5

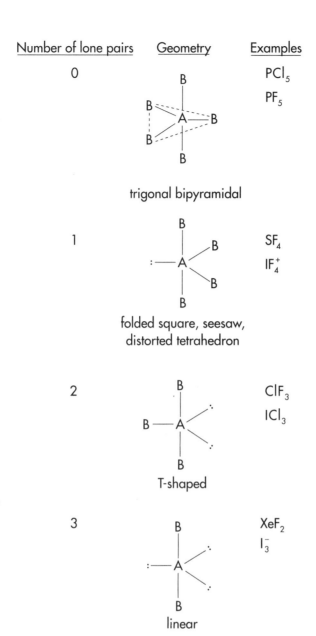

trigonal bipyramidal

| 1 | | SF_4 |
| | | IF_4^+ |

folded square, seesaw,
distorted tetrahedron

| 2 | | ClF_3 |
| | | ICl_3 |

T-shaped

| 3 | | XeF_2 |
| | | I_3^- |

linear

In trigonal bipyrimidal shapes, place the lone pairs in axial position first. In octahedral shapes, place lone pairs in equatorial position first.

If the central atom has **6** electron pairs, then it has d^2sp^3 hybridization and its basic shape is **octahedral.**

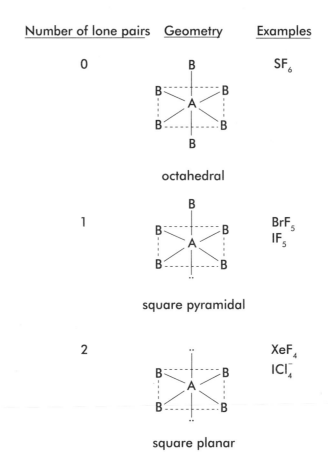

Number of lone pairs	Geometry	Examples
0	octahedral	SF_6
1	square pyramidal	BrF_5 IF_5
2	square planar	XeF_4 ICl_4^-

Bonding and Phases

The phase of a substance is directly related to the strength of their intermolecular forces. Solids have highly ordered structures where the atoms are packed tightly together, while gases has atoms spread so far apart that most of the volume is free space.

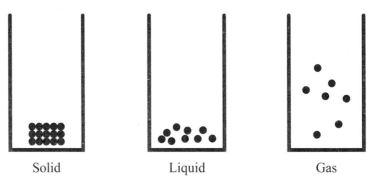

| Solid | Liquid | Gas |

Thus, the strength of the intermolecular force plays a role in the phase behavior of a substance. In other words, substances that exhibit weak interactions (e.g. London Dispersion Forces) tend to be gases at room temperature (e.g. N_2), while substances that exhibit strong interactions (e.g. ionic bonds) tend to be solids at room temperature (e.g. NaCl).

KINETIC MOLECULAR THEORY

For ideal gases, the following assumptions can be made:

- The kinetic energy of an ideal gas is directly proportional to its absolute temperature: The greater the temperature, the greater the average kinetic energy of the gas molecules.

The Average Kinetic Energy of a Single Gas Molecule

$$KE = \frac{1}{2} mv^2$$

m = mass of the molecule (kg)
v = speed of the molecule (meters/sec)
KE is measured in joules

- If several different gases are present in a sample at a given temperature, all the gases will have the same average kinetic energy. That is, the average kinetic energy of a gas depends only on the absolute temperature, not on the identity of the gas.
- The volume of an ideal gas particle is insignificant when compared with the volume in which the gas is contained.
- There are no forces of attraction between the gas molecules in an ideal gas.
- Gas molecules are in constant motion, colliding with one another and with the walls of their container.

THE IDEAL GAS EQUATION

You can use the ideal gas equation to calculate any of the four variables relating to the gas, provided that you already know the other three.

The Ideal Gas Equation

$$PV = nRT$$

P = the pressure of the gas (atm)
V = the volume of the gas (L)
n = the number of moles of gas
T = the absolute temperature of the gas (K)
R = the gas constant, 0.0821 L-atm/mol-K

You can also manipulate the ideal gas equation to figure out how changes in each of its variables affect the other variables.

$$\frac{P_1 V_1}{T_1} = \frac{P_2 V_2}{T_2}$$

P = the pressure of the gas (atm)
V = the volume of the gas (L)
T = the absolute temperature of the gas (K)

You should be comfortable with the following simple relationships:

- If the volume is constant: As pressure increases, temperature increases; as temperature increases, pressure increases.
- If the temperature is constant: As pressure increases, volume decreases; as volume increases, pressure decreases. That's Boyle's law.
- If the pressure is constant: As temperature increases, volume increases; as volume increases, temperature increases. That's Charles's law.

DALTON'S LAW

Dalton's law states that the total pressure of a mixture of gases is just the sum of all the partial pressures of the individual gases in the mixture.

Dalton's Law

$$P_{total} = P_a + P_b + P_c + \dots$$

You should also note that the partial pressure of a gas is directly proportional to the number of moles of that gas present in the mixture. So if 25 percent of the gas in a mixture is helium, then the partial pressure due to helium will be 25 percent of the total pressure.

Partial Pressure

$$P_a = (P_{total})(X_a)$$

$$X_a = \frac{\text{moles of gas A}}{\text{total moles of gas}}$$

DEVIATIONS FROM IDEAL BEHAVIOR

At low temperature and/or high pressure, gases behave in a less-than-ideal manner. That's because the assumptions made in kinetic molecular theory become invalid under conditions where gas molecules are packed too tightly together.

Two things happen when gas molecules are packed too tightly.

- *The volume of the gas molecules becomes significant.*
  The ideal gas equation does not take the volume of gas molecules into account, so the actual volume of a gas under nonideal conditions will be larger than the volume predicted by the ideal gas equation.
- *Gas molecules attract one another and stick together.*
  The ideal gas equation assumes that gas molecules never stick together. When a gas is packed tightly together, van der Waals forces (dipole–dipole attractions and London dispersion forces) become significant, causing some gas molecules to stick together. When gas molecules stick together, there are fewer particles bouncing around and creating pressure, so the real pressure in a nonideal situation will be smaller than the pressure predicted by the ideal gas equation.

DENSITY

You may be asked about the density of a gas. The density of a gas is measured in the same way as the density of a liquid or solid: in mass per unit volume.

Density of a Gas

$$D = \frac{m}{V}$$

D = density
m = mass of gas, usually in grams
V = volume occupied by a gas, usually in liters
R = the gas constant, 0.0821 L-atm/mol-K

SOLUTIONS

Molarity

Molarity (M) expresses the concentration of a solution in terms of volume. It is the most widely used unit of concentration, turning up in calculations involving equilibrium, acids and bases, and electrochemistry, among others.

When you see a chemical symbol in brackets on the test, that means they are talking about molarity. For instance, "$[Na^+]$" is the same as "the molar concentration (molarity) of sodium ions."

$$\text{Molarity } (M) = \frac{\text{moles of solute}}{\text{liters of solution}}$$

Mole Fraction

Mole fraction (X_s) gives the fraction of moles of a given substance (S) out of the total moles present in a sample. It is used in determining how the vapor pressure of a solution is lowered by the addition of a solute.

$$\text{Mole Fraction } (X_s) = \frac{\text{moles of substance S}}{\text{total number of moles in solution}}$$

Solutes and Solvents

There is a basic rule for remembering which solutes will dissolve in which solvents.

Like dissolves like

That means that polar or ionic solutes (like salt) will dissolve in polar solvents (like water). That also means that nonpolar solutes (like organic compounds) are best dissolved in nonpolar solvents. When an ionic substance dissolves, it breaks up into ions. That's **dissociation**. Free ions in a solution are called electrolytes because they can conduct electricity.

CHAPTER 4 QUESTIONS

Multiple-Choice Questions

1. Why does CaF_2 have a higher melting point than NH_3?

    (A) CaF_2 is more massive and thus has stronger London disperson forces
    (B) CaF_2 exhibits network covalent bonding, which is the strongest type of bonding
    (C) CaF_2 is smaller and exhibits greater Coulombic attractive forces
    (D) CaF_2 is an ionic substance and it requires a lot of energy to break up an ionic lattice

2. Which of the following pairs of elements is most likely to create an interstitial alloy?

    (A) titanium and copper
    (B) aluminum and lead
    (C) silver and tin
    (D) magnesium and calcium

3. Which of the following molecules would exhibit the strongest dipole moment?

    (A) CH_4
    (B) BCl_3
    (C) PBr_5
    (D) OF_2

4. Why can a molecule with the structure of NBr_5 not exist?

    (A) Nitrogen only has two energy levels, and is thus unable to expand its octet.
    (B) Bromine is much larger than nitrogen, and cannot be a terminal atom in this molecule.
    (C) It is impossible to complete the octets for all six atoms using only valence electrons.
    (D) Nitrogen does not have a low enough electronegativity to be the central atom of this molecule.

Use the following information to answer questions 5-8.

An evacuated, rigid container is filled with exactly 2.00 g of hydrogen and 4.00 g of neon. The temperature of the gases is held at 25°C and the pressure inside the container is a constant 1.0 atm.

5. What is the mole fraction of neon in the container?

    (A) 0.17
    (B) 0.33
    (C) 0.67
    (D) 0.83

6. What is the volume of the container?

    (A) 11.2 L
    (B) 22.4 L
    (C) 33.5 L
    (D) 48.8 L

7. Which gas has the higher boiling point and why?

    (A) The hydrogen, because it has a lower molar mass
    (B) The neon, because it has more electrons
    (C) The hydrogen, because it has a smaller size
    (D) The neon, because it has more protons

8. Which gas particles have a higher RMS velocity and why?

    (A) The hydrogen, because it has a lower molar mass
    (B) The neon, because it has a higher molar mass
    (C) The hydrogen, because it has a larger atomic radius
    (D) The neon, because it has a smaller atomic radius

9. Which of the following compounds would have the highest lattice energy?

    (A) LiF
    (B) $MgCl_2$
    (C) $CaBr_2$
    (D) C_2H_6

10. What is the bond order for the O-O bond in ozone, O_3?

    (A) 1.0
    (B) 1.3
    (C) 1.5
    (D) 2.0

11. A liquid whose molecules are held together by which of the following forces would be expected to have the lowest boiling point?

    (A) Ionic bonds
    (B) London dispersion forces
    (C) Hydrogen bonds
    (D) Metallic bonds

12. Which of the following species does NOT have a tetrahedral structure?

    (A) CH_4
    (B) NH_4^+
    (C) SF_4
    (D) $AlCl_4^-$

13. The six carbon atoms in a benzene molecule are shown in different resonance forms as three single bonds and three double bonds. If the length of a single carbon–carbon bond is 154 pm and the length of a double carbon–carbon bond is 133 pm, what length would be expected for the carbon–carbon bonds in benzene?

    (A) 126 pm
    (B) 133 pm
    (C) 140 pm
    (D) 154 pm

14. Which of the following could be the Lewis structure for sulfur trioxide?

(A)

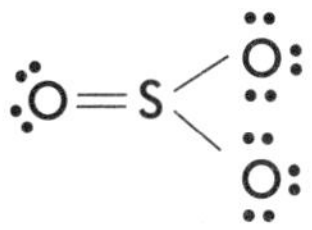

(B)

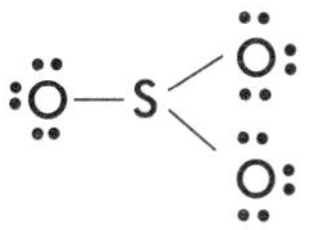

(C)

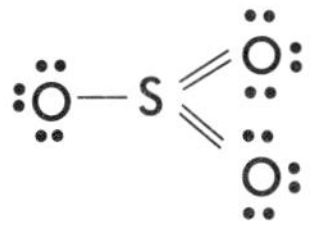

(D)

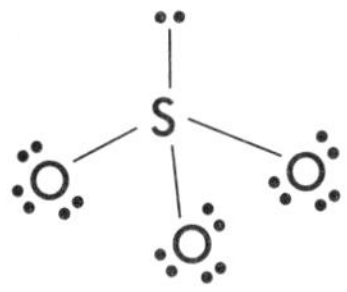

15. Which of the following molecules will have a Lewis dot structure with exactly one unshared electron pair on the central atom?

   (A) H_2O
   (B) PH_3
   (C) PCl_5
   (D) CH_2Cl_2

16. Which of the following lists of species is in order of increasing boiling points?

   (A) H_2, N_2, NH_3
   (B) N_2, NH_3, H_2
   (C) NH_3, H_2, N_2
   (D) NH_3, N_2, H_2

17. Which of the molecules listed below has the largest dipole moment?

   (A) Cl_2
   (B) HCl
   (C) SO_3
   (D) NO

18. The temperature of a sample of an ideal gas confined in a 2.0 L container was raised from 27°C to 77°C. If the initial pressure of the gas was 1,200 mmHg, what was the final pressure of the gas?

   (A)   300 mmHg
   (B)   600 mmHg
   (C) 1,400 mmHg
   (D) 2,400 mmHg

19. A sealed container containing 8.0 grams of oxygen gas and 7.0 of nitrogen gas is kept at a constant temperature and pressure. Which of the following is true?

(A) The volume occupied by oxygen is greater than the volume occupied by nitrogen.
(B) The volume occupied by oxygen is equal to the volume occupied by nitrogen.
(C) The volume occupied by nitrogen is greater than the volume occupied by oxygen.
(D) The density of nitrogen is greater than the density of oxygen.

20. A mixture of gases contains 1.5 moles of oxygen, 3.0 moles of nitrogen, and 0.5 moles of water vapor. If the total pressure is 700 mmHg, what is the partial pressure of the nitrogen gas?

(A) 210 mmHg
(B) 280 mmHg
(C) 350 mmHg
(D) 420 mmHg

21. A mixture of helium and neon gases has a total pressure of 1.2 atm. If the mixture contains twice as many moles of helium as neon, what is the partial pressure due to neon?

(A) 0.2 atm
(B) 0.3 atm
(C) 0.4 atm
(D) 0.8 atm

22. Nitrogen gas was collected over water at 25°C. If the vapor pressure of water at 25°C is 23 mmHg, and the total pressure in the container is measured at 781 mmHg, what is the partial pressure of the nitrogen gas?

(A)   46 mmHg
(B) 551 mmHg
(C) 735 mmHg
(D) 758 mmHg

23. A 22.0 gram sample of an unknown gas occupies 11.2 liters at standard temperature and pressure. Which of the following could be the identity of the gas?

(A) CO_2
(B) SO_3
(C) O_2
(D) He

24. Which of the following expressions is equal to the density of helium gas at standard temperature and pressure?

(A) $\dfrac{1}{22.4}$ g/L

(B) $\dfrac{2}{22.4}$ g/L

(C) $\dfrac{1}{4}$ g/L

(D) $\dfrac{4}{22.4}$ g/L

25. In an experiment $H_2(g)$ and $O_2(g)$ were completely reacted, above the boiling point of water, according to the following equation in a sealed container of constant volume and temperature:

$$2H_2(g) + O_2(g) \rightarrow 2H_2O(g)$$

If the initial pressure in the container before the reaction is denoted as P_i, which of the following expressions gives the final pressure, assuming ideal gas behavior?

(A) P_i
(B) $2P_i$
(C) $(3/2)P_i$
(D) $(2/3)P_i$

26. Nitrogen gas was collected over water at a temperature of 40°C, and the pressure of the sample was measured at 796 mmHg. If the vapor pressure of water at 40°C is 55 mmHg, what is the partial pressure of the nitrogen gas?

(A) 741 mmHg
(B) 756 mmHg
(C) 796 mmHg
(D) 851 mmHg

27. An ideal gas fills a balloon at a temperature of 27°C and 1 atm pressure. By what factor will the volume of the balloon change if the gas in the balloon is heated to 127°C at constant pressure?

(A) $\dfrac{27}{127}$

(B) $\dfrac{3}{4}$

(C) $\dfrac{4}{3}$

(D) $\dfrac{2}{1}$

28. A gas sample with a mass of 10 grams occupies 6.0 liters and exerts a pressure of 2.0 atm at a temperature of 26°C. Which of the following expressions is equal to the molecular mass of the gas? The gas constant, R, is 0.08 (L-atm)/(mol-K).

(A) $\dfrac{(10)(0.08)(299)}{(2.0)(6.0)}$ g/mol

(B) $\dfrac{(299)(0.08)}{(10)(2.0)(6.0)}$ g/mol

(C) $\dfrac{(2.0)(6.0)(299)}{(10)(0.08)}$ g/mol

(D) $\dfrac{(10)(2.0)(6.0)}{(299)(0.08)}$ g/mol

29. A substance is dissolved in water, forming a 0.50-molar solution. If 4.0 liters of solution contains 240 grams of the substance, what is the molecular mass of the substance?

   (A)   60 grams/mole
   (B)   120 grams/mole
   (C)   240 grams/mole
   (D)   480 grams/mole

30. A solution contains equal masses of glucose (molecular mass 180) and toluene (molecular mass 90). What is the mole fraction of glucose in the solution?

   (A) $\dfrac{1}{4}$

   (B) $\dfrac{1}{3}$

   (C) $\dfrac{1}{2}$

   (D) $\dfrac{2}{3}$

31. How many moles of Na_2SO_4 must be added to 500 milliliters of water to create a solution that has a 2-molar concentration of the Na^+ ion? (Assume the volume of the solution does not change).

   (A)   0.5 moles
   (B)   1 mole
   (C)   2 moles
   (D)   5 moles

Constructed Response Questions

1. Use the principles of bonding and molecular structure to explain the following statements.

   (a)   The boiling point of argon is −186°C, whereas the boiling point of neon is −246°C.
   (b)   Solid sodium melts at 98°C, but solid potassium melts at 64°C.
   (c)   More energy is required to break up a CaO(s) crystal into ions than to break up a KF(s) crystal into ions.
   (d)   Molten KF conducts electricity, but solid KF does not.

2. The carbonate ion CO_3^{2-} is formed when carbon dioxide, CO_2, reacts with slightly basic cold water.

   (a)   (i)   Draw the Lewis electron dot structure for the carbonate ion. Include resonance forms if they apply.
         (ii)   Draw the Lewis electron dot structure for carbon dioxide.
   (b)   Describe the hybridization of carbon in the carbonate ion.
   (c)   (i)   Describe the relative lengths of the three C–O bonds in the carbonate ion.
         (ii)   Compare the average length of the C–O bonds in the carbonate ion to the average length of the C–O bonds in carbon dioxide.

3.

Substance	Boiling Point (°C)	Bond Length (Å)	Bond Strength (kcal/mol)
H_2	−253	0.75	104.2
N_2	−196	1.10	226.8
O_2	−182	1.21	118.9
Cl_2	−34	1.99	58.0

(a) Explain the differences in the properties given in the table above for each of the following pairs.
(i) The bond strengths of N_2 and O_2
(ii) The bond lengths of H_2 and Cl_2
(iii) The boiling points of O_2 and Cl_2
(b) Use the principles of molecular bonding to explain why H_2 and O_2 are gases at room temperature, while H_2O is a liquid at room temperature.

4. H_2S, SO_4^{2-}, XeF_2, ICl_4^-

(a) Draw a Lewis electron dot diagram for each of the molecules listed above.
(b) Use the valence shell electron-pair repulsion (VSEPR) model to predict the geometry of each of the molecules.

5.

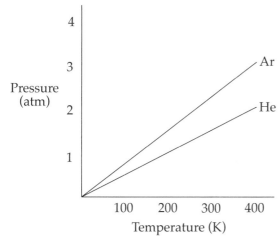

The graph above shows the changes in pressure with changing temperature of gas samples of helium and argon confined in a closed 2-liter vessel.

(a) What is the total pressure of the two gases in the container at a temperature of 200 K?
(b) How many moles of helium are contained in the vessel?
(c) How many molecules of helium are contained in the vessel?
(d) What is the ratio of the average speeds of the helium atoms to the average speeds of the argon atoms?
(e) If the volume of the container were reduced to 1 liter at a constant temperature of 300 K, what would be the new pressure of the helium gas?

6. $2 \, KClO_3(s) \rightarrow 2 \, KCl(s) + 3 \, O_2(g)$

The reaction above took place, and 1.45 liters of oxygen gas were collected over water at a temperature of 29°C and a pressure of 755 millimeters of mercury. The vapor pressure of water at 29°C is 30.0 millimeters of mercury.

(a) What is the partial pressure of the oxygen gas collected?
(b) How many moles of oxygen gas were collected?
(c) What would be the dry volume of the oxygen gas at a pressure of 760 millimeters of mercury and a temperature of 273 K?
(d) What was the mass of the $KClO_3$ consumed in the reaction?

7. Equal molar quantities of two gases, O_2 and H_2O, are confined in a closed vessel at constant temperature.

(a) Which gas, if any, has the greater partial pressure?
(b) Which gas, if any, has the greater density?
(c) Which gas, if any, has the greater concentration?
(d) Which gas, if any, has the greater average kinetic energy?
(e) Which gas, if any, will show the greater deviation from ideal behavior?

8. A student performs an experiment in which a butane lighter is held underwater directly beneath a 100-mL graduated cylinder which has been filled with water as shown in the diagram below.

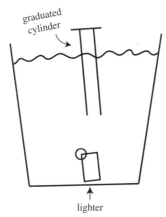

The switch on the lighter is pressed, and butane gas is released into the graduated cylinder. The student's data table for this lab is as follows:

Mass of lighter before gas release	20.432 g
Mass of lighter after gas release	20.296 g
Volume of gas collected	68.40 mL
Water Temperature	19.0°C
Atmospheric Pressure	745 mmHg

(a) Given that the vapor pressure of water at 19.0°C is 16.5 mmHg, determine the partial pressure of the butane gas collected in atmospheres.

(b) Calculate the molar mass of butane gas from the experimental data given.

(c) If the formula of butane is C_4H_{10}, determine the percent error for the student's results.

(d) The following are common potential error sources that occur during this lab. Explain whether or not each error could have been responsible for the error in the student's results.

   (i) The lighter was not sufficiently dried before massing is after the gas was released.

   (ii) The gas in the lighter was not held underwater long enough to sufficiently cool it to the same temperature of the water and was actually at a higher temperature than the water.

   (iii) Not all of the butane gas released was collected in the graduated cylinder.

CHAPTER 4 ANSWERS AND EXPLANATIONS

Multiple-Choice Questions

1. **D** Calcium is a metal, and fluorine is a nonmetal. Their electronegativities are sufficiently different for them to create an ionic bond, which is stronger than all other bond types except for network covalent bonding (which NH_3 does not exhibit)

2. **B** Interstitial alloys form when atoms of greatly different sizes combine. The aluminum atoms would have a chance to fit in between the comparatively larger lead atoms.

3. **D** When drawing the Lewis diagrams for the four choices, the first three are completely non-polar. Only OF_2, with a bent shape, would exhibit any dipoles.

non polar   non polar   non polar

4. **A** Only atoms with at least three energy levels ($n = 3$ and above) have empty d-orbitals that additional electrons can fit into, thus expanding their octet.

5. **C** Moles of $H_2 = \dfrac{2.00 \text{ g}}{2.00 \text{ g/mol}} = 1.00$ mol (remember, hydrogen is a diatomic)

    Moles of $Ne = \dfrac{4.00 \text{ g}}{20.0 \text{ g/mol}} = 0.500$ mol

    Total moles = 1.50 moles. $X_{Ne} = \dfrac{\text{moles Ne}}{\text{total moles}} = \dfrac{0.500}{1.500} = 0.67$

6. **C** At STP, 1 mol of gas takes up 22.4 L of space: $(1.5)(22.4) = 33.5$ L

7. **B** Both gases only have London dispersion forces. The more electrons a gas has, the more polarizable it is and the stronger the intermolecular forces are.

8. **A** The gas molecules have the same amount of kinetic energy due to their temperature being the same. Via $KE = \dfrac{1}{2}mv^2$, if KE is the same the molecule with less mass must correspondingly have a higher velocity.

9. **B** First, C_2H_6 is not an ionic substance and thus has no lattice energy. Next, LiF is composed of ions with charges +1 and –1, and will not be as strong as the two compounds which have ions with charges of +2 and –1. Finally, $MgCl_2$ is smaller than $CaBr_2$, meaning it will have a higher lattice energy, as according to Coulomb's law atomic radius is inversely proportional with bond energy.

10. **C** Ozone exhibits two resonance structures, shown below. Taking either bond position yields a bond order of $\frac{(1+2)}{2} = 1.5$

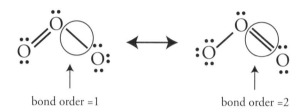

bond order =1   bond order =2

11. **B** A liquid with a low boiling point must be held together by weak bonds. London dispersion forces are the weakest kind of intermolecular forces.

12. **C** SF_4 has 34 valence electrons distributed in the Lewis dot structure and shape shown below.

In this molecule, sulfur forms dsp^3 hybrid orbitals, which have a trigonal bipyramid structure. Because SF_4 has one unshared electron pair, the molecule takes the "seesaw" or "folded square" shape.

Choices (A) and (B), CH_4 and NH_4^+, each have 8 valence electrons distributed in the same Lewis dot structure and shape, shown below for NH_4^+.

$$\left[\begin{array}{c} H \\ | \\ H\diagdown N \diagup H \\ | \\ H \end{array} \right]^+$$

In these molecules, the central atom forms sp^3 hybrid orbitals, which have a tetrahedral structure. There are no unshared electron pairs on the central atom, so the molecules are tetrahedral.

Choice (D) $AlCl_4^-$ has 32 valence electrons distributed in the same Lewis dot structure and shape, shown below for $AlCl_4^-$.

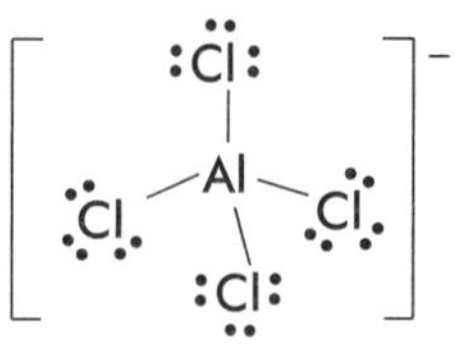

In these molecules, the central atom forms sp^3 hybrid orbitals, which have a tetrahedral structure. There are no unshared electron pairs on the central atom, so the molecules are tetrahedral.

13. **C** Resonance is used to describe a situation that lies between single and double bonds, so the bond length would also be expected to be in between that of single and double bonds.

14. **A** Choice (A) has the correct number of valence electrons $(6 + 6 + 6 + 4 + 2 = 24)$, and 8 valence electrons on each atom.

About the other choices:

(B) There are only 6 valence electrons on the sulfur atom.

(C) There are too many valence electrons (10) on the sulfur atom.

(D) There are too many valence electrons (26).

15.   **B**   The Lewis dot structures for the answer choices are shown below. Only PH_3 has a single unshared electron pair on its central atom.

(A)

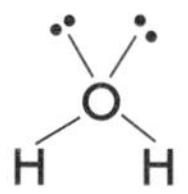

(B)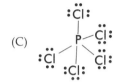

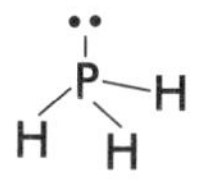

(C)

(D)

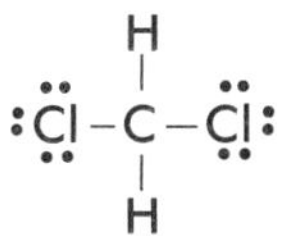

16.   **A** H_2 experiences only van der Waals forces and has the lowest boiling point.

N$_2$ also experiences only van der Waals forces, but it is larger than H_2 and has more electrons, so it has stronger van der Waals interactions with other molecules.

NH_3 is polar and undergoes hydrogen bonding, so it has the strongest intermolecular interactions and the highest boiling point.

17.   **B**   The bond that holds HCl together is a covalent bond with a large polarity. The bond that holds NO together is also polar covalent, but its polarity is very small because N and O are so close together on the periodic table. Cl_2 and N_2 are nonpolar because they share electrons equally, and SO_3 is nonpolar because it is symmetrical (trigonal planar). Nonpolar molecules have dipole moments of zero.

18. **C**   Remember to convert Celsius to Kelvin (°C + 273 = K).

27°C = 300 K and 77°C = 350 K.

From the relationship $\dfrac{P_1}{T_1} = \dfrac{P_2}{T_2}$ we get

$$\frac{(1,200 \text{ mm Hg})}{(300 \text{ K})} = \frac{P_2}{(350 \text{ K})}$$

So P_2 = 1,400 mmHg

19. **B**   8.0 grams of oxygen and 7.0 grams of nitrogen both equal 0.25 moles. Avogadro's law states equal volumes contain equal moles. This makes (A) and (C) wrong.

(D) This is reversed: The density of oxygen gas would be greater than the density of nitrogen.

20. **D**   From Dalton's law, the partial pressure of a gas depends on the number of moles of the gas that are present.

The total number of moles of gas present is

$$1.5 + 3.0 + 0.5 = 5.0 \text{ total moles}$$

If there are 3 moles of nitrogen, then $\dfrac{3}{5}$ of the pressure must be due to nitrogen.

$$(\tfrac{3}{5})(700 \text{ mmHg}) = 420 \text{ mmHg}$$

21. **C**   From Dalton's law, the partial pressure of a gas depends on the number of moles of the gas that are present. If the mixture has twice as many moles of helium as neon, then the mixture must be $\dfrac{1}{3}$ neon. So $\dfrac{1}{3}$ of the pressure must be due to neon.

$$(\tfrac{1}{3})(1.2 \text{ atm}) = 0.4 \text{ atm.}$$

22. **D**   From Dalton's law, the partial pressures of nitrogen and water vapor must add up to the total pressure in the container. The partial pressure of water vapor in a closed container will be equal to the vapor pressure of water, so the partial pressure of nitrogen is

$$781 \text{ mmHg} - 23 \text{ mmHg} = 758 \text{ mmHg}$$

23. **A** Use the following relationship:

$$\text{Moles} = \frac{\text{liters}}{22.4 \text{ L/mol}}$$

$$\text{Moles of unknown gas} = \frac{11.2 \text{ L}}{22.4 \text{ L/mol}} = 0.500 \text{ moles}$$

$$\text{MW} = \frac{\text{grams}}{\text{mole}}$$

$$\text{MW of unknown gas} = \frac{22.0 \text{ g}}{0.500 \text{ mole}} = 44.0 \text{ grams/mole}$$

That's the molecular weight of CO_2.

24. **D** Density is measured in grams per liter. One mole of helium gas has a mass of 4 grams and occupies a volume of 22.4 liters at STP, so the density of helium gas at STP is $\frac{4}{22.4}$ g/L.

25. **D** As long as the gases act ideally, the total pressure in the container will be a function only of the number of gas moles present so long as volume and temperature are held constant. The relationship between moles and pressure before and after the reaction, from the ideal gas equation ($PV = nRT$), will be $P_i/n_i = P_f/n_f$ where n_i, P_f, n_f are the initial moles, the final pressure, and the final moles respectively. We can rearrange this equation to $P_f = P_i(n_f/n_i)$, and since there are 2 moles of gas in the products for every 3 moles in the reactants, we can say that $P_f = (2/3)P_i$.

26. **A** From Dalton's law, the partial pressures of nitrogen and water vapor must add up to the total pressure in the container. The partial pressure of water vapor in a closed container will be equal to its vapor pressure, so the partial pressure of nitrogen is

796 mmHg – 55 mmHg = 741 mmHg

27. **C** From the ideal gas laws, for a gas sample at constant pressure

$$\frac{V_1}{T_1} = \frac{V_2}{T_2}$$

Solving for V_2 we get $V_2 = V_1 \frac{T_2}{T_1}$

So V_1 is multiplied by a factor of $\frac{T_2}{T_1}$

Remember to convert Celsius to Kelvin, $\frac{127\,°C + 273}{27\,°C + 273} = \frac{400 \text{ K}}{300 \text{ K}} = \frac{4}{3}$

28. **A** We can find the number of moles of gas from $PV = nRT$.

Remember to convert 26°C to 299 K.

Then solve for n.

$$n = \frac{PV}{RT} = \frac{(2.0)(6.0)}{(0.08)(299)} \text{ mol}$$

Now, remember:

$$\text{MW} = \frac{\text{grams}}{\text{moles}} = \frac{(10\text{ g})}{\left(\dfrac{(2.0)(6.0)}{(0.08)(299)}\text{ mol}\right)} = \frac{(10)(0.08)(299)}{(2.0)(6.0)} \text{ g/mol}$$

29. **B** First find the number of moles.

Moles = (molarity)(volume)
Moles of substance = (0.50 M)(4.0 L) = 2 moles

$$\text{Moles} = \frac{\text{grams}}{\text{MW}}$$

$$\text{So MW} = \frac{240\text{ g}}{2\text{ mol}} = 120 \text{ g/mol}$$

30. **B** Let's say the solution contains 180 grams of glucose and 180 grams of toluene. That's 1 mole of glucose and 2 moles of toluene. So that's 1 mole of glucose out of a total of 3 moles, for a mole fraction of $\frac{1}{3}$.

31. **A** Let's find out how many moles of Na^+ we have to add.

Moles = (molarity)(volume)
Moles of Na^+ = (2 M)(0.5 L) = 1 mole

Because we get 2 moles of Na^+ ions for every mole of Na_2SO_4 we add, we need to add only 0.5 moles of Na_2SO_4.

Constructed Response Questions

1. (a) Molecules of noble gases in the liquid phase are held together by London dispersion forces, which are weak interactions brought about by instantaneous polarities in nonpolar atoms and molecules.

Atoms with more electrons are more easily polarized and experience stronger London dispersion forces. Argon has more electrons than neon, so it experiences stronger London dispersion forces and boils at a higher temperature.

(b) Sodium and potassium are held together by metallic bonds and positively charged ions in a delocalized sea of electrons.

Potassium is larger than sodium, so the electrostatic attractions that hold the atoms together act at a greater distance, reducing the attractive force and resulting in its lower melting point.

(c) Both CaO(s) and KF(s) are held together by ionic bonds in crystal lattices.

Ionic bonds are held together by an electrostatic force, which can be determined by using Coulomb's law.

$$F = k\frac{Q_1 Q_2}{r^2}$$

CaO is more highly charged, with Ca^{2+} bonded to O^{2-}. So for CaO, Q_1 and Q_2 are +2 and –2.

KF is not as highly charged, with K^+ bonded to F^-. So for KF, Q_1 and Q_2 are +1 and –1.

CaO is held together by stronger forces and is more difficult to break apart.

(d) KF is composed of K^+ and F^- ions. In the liquid (molten) state, these ions are free to move and can thus conduct electricity.

In the solid state, the K^+ and F^- ions are fixed in a crystal lattice and their electrons are localized around them, so there is no charge that is free to move and thus no conduction of electricity.

2. (a) (i)

(ii)

(b) The central carbon atom forms three sigma bonds with oxygen atoms and has no free electron pairs, so its hybridization must be sp^2.

(c) (i) All three bonds will be the same length because no particular resonance form is preferred over the others. The actual structure is an average of the resonance structures.

(ii) The C–O bonds in the carbonate ion have resonance forms between single and double bonds, while the C–O bonds in carbon dioxide are both double bonds.

The bonds in the carbonate ion will be shorter than single bonds and longer than double bonds, so the carbonate bonds will be longer than the carbon dioxide bonds.

3. (a) (i) The bond strength of N_2 is larger than the bond strength of O_2 because N_2 molecules have triple bonds and O_2 molecules have double bonds. Triple bonds are stronger and shorter than double bonds.

(ii) The bond length of H_2 is smaller than the bond length of Cl_2 because hydrogen is a smaller atom than chlorine, allowing the hydrogen nuclei to be closer together.

(iii) Liquid oxygen and liquid chlorine are both nonpolar substances that experience only London dispersion forces of attraction. These forces are greater for Cl_2 because it has more electrons (which makes it more polarizable), so Cl_2 has a higher boiling point than O_2.

(b) H_2 and O_2 are both nonpolar molecules that experience only London dispersion forces, which are too weak to form the bonds required for a substance to be liquid at room temperature.

H_2O is a polar substance whose molecules form hydrogen bonds with each other. Hydrogen bonds are strong enough to form the bonds required in a liquid at room temperature.

4. (a)

(b) H_2S has two bonds and two free electron pairs on the central S atom. The greatest distance between the electron pairs is achieved by tetrahedral arrangement. The electron pairs at two of the four corners will cause the molecule to have a bent shape, like water.

SO_4^{2-} has four bonds around the central S atom and no free electron pairs. The four bonded pairs will be farthest apart when they are arranged in a tetrahedral shape, so the molecule is tetrahedral.

XeF_2 has two bonds and three free electron pairs on the central Xe atom. The greatest distance between the electron pairs can be achieved by a trigonal bipyramidal arrangement. The three free electron pairs will occupy the equatorial positions, which are 120 degrees apart, to minimize repulsion. The two F atoms are at the poles, so the molecule is linear.

ICl_4^{-} has four bonds and two free electron pairs on the central I atom. The greatest distance between the electron pairs can be achieved by an octahedral arrangement. The two free electron pairs will be opposite each other to minimize repulsion. The four Cl atoms are in the equatorial positions, so the molecule is square planar.

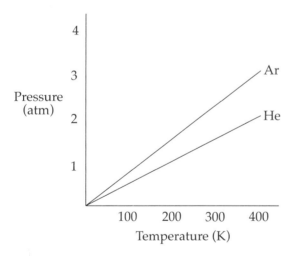

5. (a) Read the graph, and add the two pressures.

$$P_{Total} = P_{He} + P_{Ar}$$

$$P_{Total} = (1 \text{ atm}) + (1.5 \text{ atm}) = 2.5 \text{ atm}$$

(b) Read the pressure (1 atm) at 200 K, and use the ideal gas equation.

$$n = \frac{PV}{RT} = \frac{(1.0 \text{ atm})(2.0 \text{ L})}{(0.082 \text{ L-atm/mol-K})(200 \text{ K})} = 0.12 \text{ moles}$$

(c) Use the definition of a mole.

Molecules = (moles)(6.02×10^{23})

Molecules (atoms) of helium = $(0.12)(6.02 \times 10^{23}) = 7.2 \times 10^{22}$

(d) Use Graham's law.

$$\frac{v_1}{v_2} = \sqrt{\frac{MW_2}{MW_1}}$$

$$\frac{v_{He}}{v_{Ar}} = \sqrt{\frac{MW_{Ar}}{MW_{He}}} = \sqrt{\frac{(40)}{(4)}} = 3.2 \text{ to } 1$$

(e) Use the relationship

$$\frac{P_1V_1}{T_1} = \frac{P_2V_2}{T_2}$$

Since T is a constant, the equation becomes:

$$P_1V_1 = P_2V_2$$

$$(1.5 \text{ atm})(2.0 \text{ L}) = P_2(1.0 \text{ L})$$

$$P_2 = 3.0 \text{ atm}$$

6. (a) Use Dalton's law.

$$P_{Total} = P_{Oxygen} + P_{Water}$$

$$(755 \text{ mmHg}) = (P_{Oxygen}) + (30.0 \text{ mmHg})$$

$$P_{Oxygen} = 725 \text{ mmHg}$$

(b) Use the ideal gas law. Don't forget to convert to the proper units.

$$n = \frac{PV}{RT} = \frac{\left(\frac{725}{760} \text{ atm}\right)(1.45 \text{ L})}{(0.082 \text{ L-atm/mol-K})(302 \text{ K})} = 0.056 \text{ moles}$$

(c) At STP, moles of gas and volume are directly related.

Volume = (moles)(22.4 L/mol)

Volume of O_2 = (0.056 mol)(22.4 L/mol) = 1.25 L

(d) We know that 0.056 moles of O_2 were produced in the reaction.

From the balanced equation, we know that for every 3 moles of O_2 produced, 2 moles of $KClO_3$ are consumed. So there are $\frac{2}{3}$ as many moles of $KClO_3$ as O_2.

Moles of $KClO_3$ = $(\frac{2}{3})$(moles of O_2)

Moles of $KClO_3$ = $(\frac{2}{3})$(0.056 mol) = 0.037 moles

Grams = (moles)(MW)

Grams of $KClO_3$ = (0.037 mol)(122 g/mol) = 4.51 g

7. (a) The partial pressures depend on the number of moles of gas present. Because the number of moles of the two gases are the same, the partial pressures are the same.

(b) O_2 has the greater density. Density is mass per unit volume. Both gases have the same number of moles in the same volume, but oxygen has heavier molecules, so it has greater density.

(c) Concentration is moles per volume. Both gases have the same number of moles in the same volume, so their concentrations are the same.

(d) According to kinetic-molecular theory, the average kinetic energy of a gas depends only on the temperature. Both gases are at the same temperature, so they have the same average kinetic energy.

(e) H_2O will deviate most from ideal behavior. Ideal behavior for gas molecules assumes that there will be no intermolecular interactions.

H_2O is polar and O_2 is not. H_2O undergoes hydrogen bonding while O_2 does not. So H_2O has stronger intermolecular interactions, which will cause it to deviate more from ideal behavior.

8. (a) 745 mmHg – 16.5 mmHg = 729 mmHg

$\frac{729 \text{ mmHg}}{760 \text{ mmHg}} = 0.959$ atm

(b) To determine the mass of the butane, subtract the mass of the lighter after the butane was released from the mass of the lighter before the butane was released.

20.432 g – 20.296 g = 0.136 g

To determine the moles of butane, use the ideal gas law, making any necessary conversions first.

$$PV = nRT$$
$$(0.959 \text{ atm})(0.06840 \text{ L}) = n(0.0821 \text{ atm*L/mol*K})(292 \text{ K})$$
$$n = 2.74 \times 10^{-3} \text{ mol}$$

Molar mass is defined as grams per mole, so
$0.136 \text{ g}/2.74 \times 10^{-3} \text{ mol} = 49.6 \text{ g/mol}$

(c) Actual molar mass of butane:

$(12.00 \text{ g/mol} \times 4) + (1.01 \text{ g/mol} \times 10) = 58.08 \text{ g/mol}$

Percent error is:

$$\frac{|\text{Actual value} - \text{experimental value}|}{\text{Actual value}} \times 100\%$$

So:

$$\frac{|58.08 - 49.6|}{58.08} \times 100\% = 14.5\% \text{ error}$$

(d) (i) If the lighter is not sufficiently dried, then the mass of the butane calculated will be artificially low. That means the numerator in the molar mass calculation will be too low, which would lead to an experimental molar mass that is too low. This is consistent with the student's error.

(ii) If the temperature of the butane is higher than the water temperature, the calculated moles of butane will be artificially high. This means the denominator in the molar mass calculation will be too high, which would lead to an experimental molar mass that is too low. This is consistent with the student's error.

(iii) If some butane gas escaped without going into the graduated cylinder, the volume of butane gas collected will be artificially low. That will make the calculation for moles of butane too low, which in turns means the denominator of the molar mass calculation will be too low. This would lead to an experimental molar mass that is too high. This is NOT consistent with the student's error.

Chapter 5
Big Idea #3: Chemical Reactions, Energy Changes, and Redox Reactions

Changes in matter involve the rearrangement and/or reorganization of atoms and/or the transfer of electrons

TYPES OF REACTIONS

Reactions may be classified into several categories.

1. Synthesis Reactions – When simple compounds are combined to form a single, more complex compound.

   $$2\,Mg(s) + O_2(g) \rightarrow 2\,MgO(s)$$

2. Decomposition – The opposite of a synthesis. A reaction where a single compound is split into two or more simple compounds, usually in the presence of heat.

   $$HgO(s) + Heat \rightarrow Hg(s) + \tfrac{1}{2}\,O_2(g)$$

3. Acid-Base Reaction – A reaction when an acid (i.e. H^+) reacts with a base (i.e. OH^-) to form water and a salt.

   $$HCl(aq) + NaOH(aq) \rightarrow NaCl(aq) + H_2O(l)$$

4. Oxidation-Reduction (Redox) Reaction – A reaction that results in the change of the oxidation states of some participating molecules.

   $$Cu^{2+}(aq) + 2e^- \rightarrow Cu(s)$$

5. Precipitation – When two aqueous solutions mix, sometimes a new cation/anion pairing can create an insoluble salt. This type of reaction is called a precipitation reaction. When potassium carbonate and magnesium nitrate mix, a precipitate of magnesium carbonate will form as follows:

   $$K_2CO_3\,(aq) + Mg(NO_3)_2\,(aq) \rightarrow 2\,KNO_3\,(aq) + MgCO_3\,(s)$$

   That can also be written as a net ionic equation. In solution, the potassium and nitrate ions are do not actually take part in the reaction. They start out as free ions, and end up as free ions. We call those ions **spectator ions**. The only thing that is changing is that carbonate and magnesium ions are bonding to form magnesium carbonate. The net ionic equation is:

   $$CO_3^{2-} + Mg^{2+} \rightarrow MgCO_3\,(s)$$

   Finally, this can be represented using particulate diagrams, as shown below:

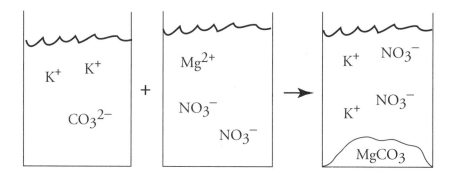

You may have heard that you need to memorize solubility rules—that is, what ions will form insoluble precipitates when you combine. You DO NOT need to memorize most of these; for the most part the AP exam will provide you with them as needed. The only ones you need to know are:

1. Compounds with an alkali metal cation (Na^+, Li^+, K^+, etc) or an ammonium cation (NH_4^+) are always soluble.
2. Compounds with a nitrate (NO_3^-) anion are always soluble.

That's it. You should know how to interpret any solubility rules that you are given on the test, but you need not memorize anything beyond those two.

Ionic substances that dissolve in water do so because the attraction of the ions to the dipoles of the water molecules. Let's take NaBr for example:

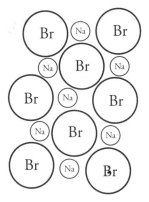

When it dissolves in water, the positive Na^+ cations are attracted to the negative (oxygen) ends of the water molecules. The negative Br^- anions are attracted to the positive (hydrogen) ends of the water molecules. Thus, when the substance dissolves it looks like this on the particulate level:

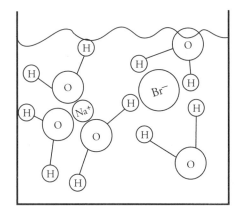

CHEMICAL EQUATIONS

Balancing Chemical Equations

Normally, balancing a chemical equation is a trial-and-error process. You start with the most complicated-looking compound in the equation and work from there. There is, however, an old Princeton Review SAT trick that you may want to try if you see a balancing equation question on the multiple-choice section. The trick is called **backsolving**.

It works like this: To make a balancing equation question work in a multiple-choice format, one of the answer choices is the correct coefficient for one of the species in the reaction. So instead of starting blind in the trial-and-error process, you can insert the answer choices one by one to see which one works. You probably won't have to try all five, and if you start in the middle and the number doesn't work, it might be obviously too small or large, eliminating other choices before you have to try them. Let's try it.

$$...NH_3 + ...O_2 \rightarrow ...N_2 + ...H_2O$$

1. If the equation above were balanced with lowest whole number coefficients, the coefficient for NH_3 would be

   (A) 1
   (B) 2
   (C) 3
   (D) 4

Here's How to Crack It!

Start at (C) because it's the middle number. If there are 3 NH_3's, then there can't be a whole number coefficient for N_2, so (C) is wrong, and so is the other odd number answer, (A).

Try (D).

If there are 4 NH_3's, then there must be 2 N_2's and 6 H_2O's.

If there are 6 H_2O's, then there must be 3 O_2's, and the equation is balanced with lowest whole number coefficients.

Backsolving is more efficient than the methods that you're accustomed to. If you use the answer choices that you're given, you streamline the trial-and-error process and allow yourself to use POE as you work on the problem.

Chemical Equations and Calculations

Many of the stoichiometry problems on the test will be formatted in the following way: You will be given a balanced chemical equation and told that you have some number of grams (or liters of gas, or molar concentration, and so on) of reactant. Then you will be asked what number of grams (or liters of gas, or molar concentration, and so on) of products are generated.

In these cases, follow this simple series of steps.

1. Convert whatever quantity you are given into moles.
2. If you are given information about two reactants, you may have to use the equation coefficients to determine which one is the limiting reagent. Remember: The limiting reagent is not necessarily the reactant that you have the least of; it is the reactant that runs out first.
3. Use the balanced equation to determine how many moles of the desired product are generated.
4. Convert moles of product to the desired unit.

Let's try one.

$$2 \text{ HBr}(aq) + \text{Zn}(s) \rightarrow \text{ZnBr}_2(aq) + \text{H}_2(g)$$

2. A piece of solid zinc weighing 98 grams was added to a solution containing 324 grams of HBr. What is the volume of H_2 produced at standard temperature and pressure if the reaction above runs to completion?

   (A) 11 liters
   (B) 22 liters
   (C) 34 liters
   (D) 45 liters

Here's How to Crack It!

1. Convert whatever quantity you are given into moles.

$$\text{Moles of Zn} = \frac{\text{grams}}{\text{molar mass}} = \frac{(98 \text{ g})}{(65.4 \text{ g/mol})} = 1.5 \text{ mol}$$

$$\text{Moles of HBr} = \frac{\text{grams}}{\text{molar mass}} = \frac{(324 \text{ g})}{(80.9 \text{ g/mol})} = 4.0 \text{ mol}$$

2. Use the balanced equation to find the limiting reagent.
   From the balanced equation, 2 moles of HBr are used for every mole of Zn that reacts, so when 1.5 moles of Zn react, 3 moles of HBr are consumed, and there will be HBr left over when all of the Zn is gone. That makes Zn the limiting reagent.
3. Use the balanced equation to determine how many moles of the desired product are generated.
   1 mole of H_2 is produced for every mole of Zn consumed, so if 1.5 moles of Zn are consumed, then 1.5 moles of H_2 are produced.
4. Convert moles of product to the desired unit.
   The H_2 gas is at STP, so we can convert directly from moles to volume.
   Volume of H_2 = (moles)(22.4 L/mol) = (1.5 mol)(22.4 L/mol) = 33.6 L $\approx$ 34 L

So (C) is correct.

———————————— ◯ ————————————

Let's try another one using the same reaction.

———————————— ◯ ————————————

$$2\ HBr(aq) + Zn(s) \rightarrow ZnBr_2(aq) + H_2(g)$$

3. A piece of solid zinc weighing 13.1 grams was placed in a container. A 0.10-molar solution of HBr was slowly added to the container until the zinc was completely dissolved. What was the volume of HBr solution required to completely dissolve the solid zinc?

   (A) 1.0 L
   (B) 2.0 L
   (C) 3.0 L
   (D) 4.0 L

Here's How to Crack It!

1. Convert whatever quantity you are given into moles.

$$\text{Moles of Zn} = \frac{\text{grams}}{\text{MW}} = \frac{(13.1\ g)}{(65.4\ g/mol)} = 0.200\ mol$$

2. Use the balanced equation to find the limiting reagent.
3. Use the balanced equation to determine how many moles of the desired product are generated.
   In this case, we're using the balanced reaction to find out how much of one reactant is required to consume the other reactant. It's a slight variation on the process described in (2) and (3).
   We can see from the balanced equation that it takes 2 moles of HBr to react completely with 1 mole of Zn, so it will take 0.400 moles of HBr to react completely with 0.200 moles of Zn.

4. Convert moles of product to the desired unit.
   Moles of HBr = (molarity)(volume)

$$\text{Volume of HBr} = \frac{\text{moles}}{\text{molarity}} = \frac{(0.400\,\text{mol})}{(0.10\,\text{mol/L})} = 4.0\text{ L}$$

So (D) is correct.

When you perform calculations, always include units. Including units in your calculations will help you (and the person scoring your test) keep track of what you are doing. Including units will also get you partial credit points on the free-response section.

ENTHALPY

Enthalpy Change, ΔH

The enthalpy of a substance is a measure of the energy that is released or absorbed by the substance when bonds are broken and formed during a reaction.

> ### The Basic Rules of Enthalpy
> When bonds are *formed*, energy is *released*.
> When bonds are *broken*, energy is *absorbed*.

The change in enthalpy, ΔH, that takes place over the course of a reaction can be calculated by subtracting the enthalpy of the reactants from the enthalpy of the products.

> ### Enthalpy Change
> $$\Delta H = H_{\text{products}} - H_{\text{reactants}}$$

If the products have stronger bonds than the reactants, then the products have lower enthalpy than the reactants and are more stable; in this case, energy is released by the reaction, or the reaction is **exothermic.**

If the products have weaker bonds than the reactants, then the products have higher enthalpy than the reactants and are less stable; in this case, energy is absorbed by the reaction, or the reaction is **endothermic.**

All substances like to be in the lowest possible energy state, which gives them the greatest stability. This means that, in general, exothermic processes are more likely to occur spontaneously than endothermic processes.

ENERGY DIAGRAMS

Exothermic and Endothermic Reactions

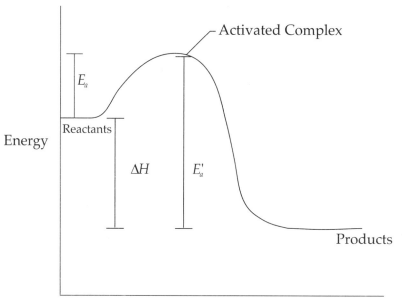

EXOTHERMIC REACTION

The diagram above shows the energy change that takes place during an exothermic reaction. The reactants start with a certain amount of energy (read the graph from left to right). For the reaction to proceed, the reactants must have enough energy to reach the transition state, where they are part of an activated complex. This is the highest point on the graph above. The amount of energy needed to reach this point is called the **activation energy, E_a**. At this point, all reactant bonds have been broken, but no product bonds have been formed, so this is the point in the reaction with the highest energy and lowest stability. The energy needed for the reverse reaction is shown as line E_a'.

Moving to the right past the activated complex, product bonds start to form, and we eventually reach the energy level of the products.

This diagram represents an exothermic reaction, so the products are at a lower energy level than the reactants and ΔH is negative.

The diagram below shows an endothermic reaction.

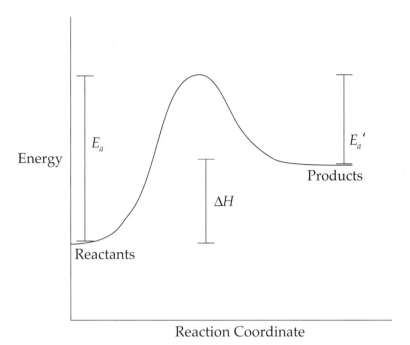

ENDOTHERMIC REACTION

In this diagram, the energy of the products is greater than the energy of the reactants, so ΔH is positive.

Reaction diagrams can be read in both directions, so the reverse reaction for an exothermic reaction is endothermic and vice versa.

CATALYSTS AND ENERGY DIAGRAMS

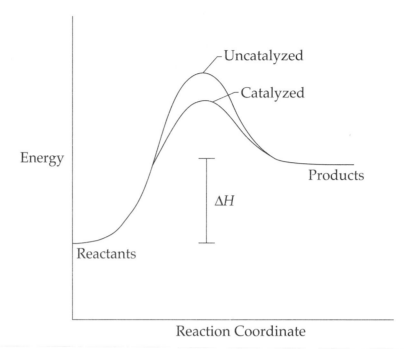

A catalyst speeds up a reaction by providing the reactants with an alternate pathway that has a lower activation energy, as shown in the diagram above.

Notice that the only difference between the catalyzed reaction and the uncatalyzed reaction is that the energy of the activated complex is lower for the catalyzed reaction. A catalyst lowers the activation energy, but it has no effect on the energy of the reactants, the energy of the products, or ΔH for the reaction.

Also note that a catalyst lowers the activation energy for both the forward and the reverse reaction, so it has no effect on the equilibrium conditions.

OXIDATION STATES

The **oxidation state** (or oxidation number) of an atom indicates the number of electrons that it gains or loses when it forms a bond. For instance, upon forming a bond with another atom, oxygen generally gains two electrons, which are negatively charged, so the oxidation state of oxygen in a bond is –2. Similarly, sodium generally loses one electron when it bonds to another atom, so its oxidation state in a bond is +1. Here are three important things you have to keep in mind when dealing with oxidation numbers.

- The oxidation state of an atom that is not bonded to an atom of another element is zero. That means either an atom that is not bonded to any other atom or an atom that is bonded to another atom of the same element (like the oxygen atoms in O_2).
- The oxidation numbers for all the atoms in a molecule must add up to zero.

- The oxidation numbers for all the atoms in a polyatomic ion must add up to the charge on the ion.

Most elements have different oxidation numbers that can vary depending on the molecule that they are a part of. The following chart shows some elements that consistently take the same oxidation numbers as their corresponding molecules.

Element	Oxidation Number
Alkali metals (Li, Na, ...)	+1
Alkaline earths (Be, Mg, ...)	+2
Group 3A (B, Al, ...)	+3
Oxygen	−2
Halogens (F, Cl, ...)	−1

Transition metals can have several oxidation states, which are differentiated from one another by a Roman numeral in the name of the compound. For example, in copper (II) sulfate ($CuSO_4$), the oxidation state for copper is +2, and in lead (II) oxide (PbO), the oxidation state for lead is +2. However, copper can also have an oxidation number of +1, and Pb can range from −4 to +4. In general, transition metals or d-block metals are characterized by variable or changing oxidation states.

You should be familiar with the following polyatomic ions and their charges.

Hydroxide	OH^-
Nitrate	NO_3^-
Acetate	$C_2H_3O_2^-$
Cyanide	CN^-
Permanganate	MnO_4^-
Carbonate	CO_3^{2-}
Sulfate	SO_4^{2-}
Dichromate	$Cr_2O_7^{2-}$
Phosphate	PO_4^{3-}
Ammonium	NH_4^+

OXIDATION-REDUCTION REACTIONS

In an oxidation-reduction (or redox, for short) reaction, electrons are exchanged by the reactants, and the oxidation states of some of the reactants are changed over the course of the reaction. Look at the following reaction:

$$Fe + 2\ HCl \rightarrow FeCl_2 + H_2$$

The oxidation state of Fe changes from 0 to +2.

The oxidation state of H changes from +1 to 0.

- *When an atom gains electrons, its oxidation number decreases, and it is said to have been reduced.*

In the reaction on the previous page, H was reduced.

- *When an atom loses electrons, its oxidation number increases, and it is said to have been oxidized.*

In the reaction on the previous page, Fe was oxidized.

Oxidation and reduction go hand in hand. If one atom is losing electrons, another atom must be gaining them.

> Here's a mnemonic device that might be useful.
>
> LEO the lion says GER
> **LEO:** you Lose Electrons in Oxidation
> **GER:** you Gain Electrons in Reduction

An oxidation-reduction reaction can be written as two **half-reactions**: one for the reduction and one for the oxidation. For example, the reaction

$$Fe + 2\ HCl \rightarrow FeCl_2 + H_2$$

can be written as

$$Fe \rightarrow Fe^{2+} + 2\ e^- \qquad \text{Oxidation}$$
$$2\ H^+ + 2\ e^- \rightarrow H_2 \qquad \text{Reduction}$$

Reduction Potentials

Every half-reaction has an electric potential, or voltage, associated with it. You will be given the necessary values for the standard reduction potential of half-reactions for any question in which they are required. Potentials are always given as reduction half-reactions, but you can read them in reverse and flip the sign on the voltage to get oxidation potentials.

Look at the reduction potential for Zn^{2+}.

$$Zn^{2+} + 2\ e^- \rightarrow Zn \qquad E^\circ = -0.76\ V$$

Read the reduction half-reaction in reverse and change the sign on the voltage to get the oxidation potential for Zn.

$$Zn \rightarrow Zn^{2+} + 2\ e^- \qquad E^\circ = +0.76\ V$$

The larger the potential for a half-reaction, the more likely it is to occur; for instance, let's look at the top of the table of half-reactions.

$$F_2(g) + 2e^- \rightarrow 2\ F^- \qquad E^\circ = +2.87\ V$$

$F_2(g)$ has a very large reduction potential, so it is likely to gain electrons and be reduced.

We need to look at the reverse (oxidation) reaction to get a positive potential.

$$Li(s) \rightarrow Li^+ + e^- \qquad E^0 = +3.05 \text{ V}$$

Li(s) has a very large oxidation potential, making it very likely to lose electrons and be oxidized.

You can calculate the potential of a redox reaction if you know the potentials for the two half-reactions that constitute it. There are two important things to remember when calculating the potential of a redox reaction.

- Add the potential for the oxidation half-reaction to the potential for the reduction half-reaction.
- Never multiply the potential for a half-reaction by a coefficient.

Let's look at the following reaction:

$$Zn + 2 Ag^+ \rightarrow Zn^{2+} + 2 Ag$$

The two half-reactions are

Oxidation:	$Zn \rightarrow Zn^{2+} + 2 e^-$	$E^0 = +0.76$ V
Reduction:	$Ag^+ + e^- \rightarrow Ag$	$E^0 = +0.80$ V

$$E = E_{oxidation} + E_{reduction}$$
$$E = 0.76 \text{ V} + 0.80 \text{ V} = 1.56 \text{ V}$$

Notice that we ignored that silver has a coefficient of 2 in the balanced equation.

The relative reduction strengths of two different metals can also be determined qualitatively. In the above reaction, if Zn(s) were placed in a solution containing Ag^+ ions, the silver ions have a high enough reduction potential that they would take electrons from the zinc and start forming solid silver, which would precipitate out on the surface of the zinc.

However, if Ag(s) were placed in a solution containing Zn^{2+} ions, zinc does not have a high enough reduction potential to take electrons from silver and so no reaction would occur. So, when a solid metal is placed into a metallic solution and a new solid starts to form, the reduction potential of the metal in solution is greater than that of the solid. If no solid forms, the reduction potential of the solid metal is higher.

GALVANIC CELLS

In a **galvanic cell** (also called a voltaic cell), a spontaneous redox reaction is used to generate a flow of current.

Look at the following spontaneous redox reaction:

$$Zn(s) + Cu^{2+}(aq) \rightarrow Zn^{2+}(aq) + Cu(s) \qquad E° = 1.10 \text{ V}$$

Oxidation:	$Zn(s) \rightarrow Zn^{2+}(aq) + 2 \text{ e}^-$	$E° = 0.76$ V
Reduction:	$Cu^{2+}(aq) + 2 \text{ e}^- \rightarrow Cu(s)$	$E° = 0.34$ V

In a galvanic cell, the two half-reactions take place in separate chambers, and the electrons that are released by the oxidation reaction pass through a wire to the chamber where they are consumed in the reduction reaction. That's how the current is created. **Current** is defined as the flow of positive charge, so current is always in the opposite direction from the flow of electrons.

There's a mnemonic device to remember that.

AN OX
RED CAT

In any electric cell (either a galvanic cell or an electrolytic cell, which we'll discuss a little later) oxidation takes place at the electrode called the **anode**. Reduction takes place at the electrode called the **cathode**.

The salt bridge maintains electrical neutrality in the system by providing enough negative ions to equal the positive ions being created at the anode (during oxidation) and providing positive ions to replace the Cu^{2+} ions being used up at the cathode (during reduction). The salt bridge can be an actual salt, or it can be a slim passage that allows ions to move between the two chambers.

Under standard conditions (when all concentrations are 1 M), the voltage of the cell is the same as the total voltage of the redox reaction. Under nonstandard conditions, Le Châtelier's principle can be applied to determine the voltage change. If the concentration of the products in a voltaic cell increases, the voltage decreases. If the concentration of the reactants increases, the voltage increases. This will be examined in greater detail when equilibrium is studied in Big Idea #6.

ELECTROLYTIC CELLS

In an electrolytic cell, an outside source of voltage is used to force a nonspontaneous redox reaction to take place. Most electrolytic cells occur in aqueous solutions which are created when a chemical dissolves in water. In these cases, either the ions themselves or the water itself can be reduced or oxidized.

Let's look at a solution of nickel (II) chloride as an example. To determine which substance is reduced, you must compare the reduction potential of the cation with that of water. The half-reaction with the more positive value is the one that will occur.

$$Ni^{2+} + 2e^- \rightarrow Ni \ (s) \qquad E° = -0.25 \text{ V}$$

$$2 \ H_2O + 2e^- \rightarrow H_2 \ (g) + 2 \ OH^- \qquad E° = -0.80 \text{ V}$$

In this case, the Ni^{2+} reduction will occur. For the oxidation, the oxidation potential of the anion versus that of water must be examined. As with before, the half-reaction with the most positive value is the one that will occur. Remember that potentials are always given as reduction potentials, so you must flip the sign when you flip the equation to an oxidation.

$$2\ Cl^- \rightarrow Cl_2\ (l) + 2e^- \qquad\qquad E° = -1.36\ V$$

$$2\ H_2O\ (l) \rightarrow O_2\ (g) + 4\ H^+ + 4e^- \qquad\qquad E° = -1.23\ V$$

So, in this case, the water itself would be reduced. When the equations are balanced for electron transfer, the net ionic equation looks like this:

$$2\ Ni^{2+} + 2\ H_2O\ (l) \rightarrow 2\ Ni\ (s) + O_2\ (g) + 4\ H^+$$

$$E = -0.25\ V + -1.24V = -1.49\ V$$

The anode and cathode in an electrolytic reaction are usually just metal bars that conduct current which do not take part in the reaction. In the above reaction, solid nickel is being created at the cathode, and oxygen gas is being evolved at the anode. The sign of your total cell potential (E) for an electrolytic reaction is always negative.

Occasionally, a current will either be run through a molten compound or pure water. In this case, you do not need to determine which redox reactions are taking place as you will only have one choice for each.

The AN OX/RED CAT rule applies to the electrolytic cell in the same way that it applies to the galvanic cell.

Electroplating

Electrolytic cells are used for electroplating. You may see a question on the test that gives you an electrical current and asks you how much metal "plates out."

There are roughly four steps for figuring out electrolysis problems.

1. If you know the current and the time, you can calculate the charge in coulombs.

> Current
>
> $$I = \frac{q}{t}$$
>
> I = current (amperes, A)
> q = charge (coulombs, C)
> t = time (sec)

2. Once you know the charge in coulombs, you know how many electrons were involved in the reaction.

$$\text{Moles of electrons} = \frac{\text{coulombs}}{96{,}500 \text{ coulombs/mol}}$$

3. When you know the number of moles of electrons and you know the half-reaction for the metal, you can find out how many moles of metal plated out. For example, from this half-reaction for gold

$$Au^{3+} + 3 \ e^- \rightarrow Au(s)$$

you know that for every 3 moles of electrons consumed, you get 1 mole of gold.

4. Once you know the number of moles of metal, you can use what you know from stoichiometry to calculate the number of grams of metal.

For instance, if a current of 2.50 A is run through a solution of iron (III) chloride for 15 minutes, it would cause the following mass of iron to plate out:

$$15 \text{ minutes} \times \frac{2.50 \text{ C}}{\text{minute}} \times \frac{1 \text{ mol } e^-}{96{,}500 \text{ C}} \times \frac{1 \text{ mol Fe}}{3 \text{ mol } e^-} \times \frac{55.85 \text{ g}}{1 \text{ mol Fe}} = 7.23 \times 10^{-3} \text{ g Fe}$$

It is particularly important to keep track of your units in an electroplating problem; there are four different conversions before you come up with your final answer. In general, as long as all of your conversions are set up correctly your final answer will have the correct units.

CHAPTER 5 QUESTIONS

Multiple-Choice Questions

1. A strip of metal X is placed into a solution containing Y^{2+} ions and no reaction occurs. When metal X is placed in a separate solution containing Z^{2+} ions, metal Z starts to form on the strip. Which of the following choices organizes the reduction potentials for metals X, Y, and Z from greatest to least?

    (A) X > Y > Z
    (B) Y > Z > X
    (C) Z > X > Y
    (D) Y > X > Z

Use the following solubility rules to answer questions 2 and 3:

> Salts containing halide anions are soluble except for those containing Ag^+, Pb^{2+}, and Hg_2^{2+}.
>
> Salts containing carbonate anions are insoluble except for those containing alkali metals or ammonium.

2. If solutions of iron (III) nitrate and sodium carbonate are mixed, what would be the formula of the precipitate?

    (A) Fe_3CO_3
    (B) $Fe_2(CO_3)_3$
    (C) $NaNO_3$
    (D) No precipitate would form.

3. If solutions containing equal amounts of $AgNO_3$ and KCl are mixed, what is the identity of the spectator ions?

    (A) Ag^+, NO_3^-, K^+, and Cl^-
    (B) Ag^+ and Cl^-
    (C) K^+ and Ag^+
    (D) K^+ and NO_3^-

4. Which of the following is true for an endothermic reaction?

    (A) The strength of the bonds in the products exceeds the strength of the bonds in the reactants.
    (B) The activation energy is always greater than the activation energy for an exothermic reaction.
    (C) Energy is released over the course of the reaction.
    (D) A catalyst will increase the rate of the reaction by increasing the activation energy.

5. In which of the following compounds is the oxidation number of chromium the greatest?

    (A) CrO_4^{2-}
    (B) CrO
    (C) Cr^{3+}
    (D) Cr (s)

6. On an energy diagram, the energy of the activated complex is always

    (A) between the energy levels of the products and reactants.
    (B) above the energy level of the reactants, but below that of the products.
    (C) below the energy level of the reactants, but above that of the products.
    (D) above the energy levels of both the reactants and products.

7. What is the mass ratio of fluorine to boron in a boron trifluoride molecule?

    (A) 1.8 to 1
    (B) 3.0 to 1
    (C) 3.5 to 1
    (D) 5.3 to 1

8. Based on the particulate drawing of the products for the reaction below, which reactant is limiting for the following reaction and why?

$$2 \text{ K}_3\text{PO}_4 \text{ } (aq) + 3 \text{ MgI}_2(aq) \rightarrow \text{Mg}_3(\text{PO}_4)_2 \text{ } (s) + 6 \text{ KI } (aq)$$

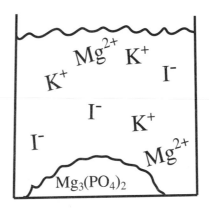

    (A) The K_3PO_4, because there are no excess PO_4^{3-} ions after the reaction.
    (B) The MgI_2, because there are excess Mg^{2+} cations remaining after the reaction.
    (C) The K_3PO_4, because it contains a cation that cannot form a precipitate.
    (D) The MgI_2, because it requires more of itself to create the products.

9. What is the mass of oxygen in 148 grams of calcium hydroxide $(Ca(OH)_2)$?

    (A) 24 grams
    (B) 32 grams
    (C) 48 grams
    (D) 64 grams

10. An ion containing only oxygen and chlorine is 31% oxygen by mass. What is its empirical formula?

    (A) ClO^-
    (B) ClO_2^-
    (C) ClO_3^-
    (D) ClO_4^-

Use the following information to answer questions 11–14

When heated in a closed container in the presence of a catalyst, potassium chlorate decomposes into potassium chloride and oxygen gas via the following reaction:

$$2KClO_3(s) \rightarrow 2KCl(s) + 3O_2(g)$$

11. If 12.25 g of potassium chlorate decomposes, how many grams of oxygen gas will be generated?

    (A) 1.60 g
    (B) 3.20 g
    (C) 4.80 g
    (D) 18.37 g

12. Approximately how many liters of oxygen gas will be evolved at STP?

    (A) 2.24 L
    (B) 3.36 L
    (C) 4.48 L
    (D) 22.4 L

13. If the temperature of the gas is doubled with the volume is held constant, what will happen to the pressure exerted by the gas and why?

    (A) It will also double as the gas molecules are hitting the sides of the container more often.
    (B) It will also double because the gas molecules are exerting a greater force on each other.
    (C) It will be cut in half because the molecules will lose more energy when colliding.
    (D) It will increase by a factor of four because the kinetic energy will be four times greater.

14. Why is a catalyst present during the reaction?

    (A) A catalyst is necessary for all decomposition reactions to occur.
    (B) A catalyst reduces the bond energy in the reactants, making them easier to activate.
    (C) A catalyst reduces the energy differential between the reactants and the products.
    (D) A catalyst lowers the activation energy of the overall reaction and speeds it up.

15. A sample of a hydrate of $CuSO_4$ with a mass of 250 grams was heated until all the water was removed. The sample was then weighed and found to have a mass of 160 grams. What is the formula for the hydrate?

    (A) $CuSO_4 \cdot 10\,H_2O$
    (B) $CuSO_4 \cdot 7\,H_2O$
    (C) $CuSO_4 \cdot 5\,H_2O$
    (D) $CuSO_4 \cdot 2\,H_2O$

16. A compound containing only sulfur and oxygen is 50% sulfur by weight. What is the empirical formula for the compound?

    (A) SO
    (B) SO_2
    (C) SO_3
    (D) S_2O

17.
$$...CN^- + ...OH^- \rightarrow ...CNO^- + ...H_2O + ...e^-$$

When the half reaction above is balanced, what is the coefficient for OH^- if all the coefficients are reduced to the lowest whole number?

(A) 1
(B) 2
(C) 3
(D) 4

18.
$$CaCO_3(s) + 2\ H+(aq) \rightarrow Ca_2+(aq) + H_2O(l) + CO_2(g)$$

If the reaction above took place at standard temperature and pressure and 150 grams of $CaCO_3(s)$ were consumed, what was the volume of $CO_2(g)$ produced at STP?

(A) 11 L
(B) 22 L
(C) 34 L
(D) 45 L

19. A gaseous mixture at 25°C contained 1 mole of CH_4 and 2 moles of O_2 and the pressure was measured at 2 atm. The gases then underwent the reaction shown below.

$$CH_4(g) + 2\ O_2(g) \rightarrow CO_2(g) + 2\ H_2O(g)$$

What was the pressure in the container after the reaction had gone to completion and the temperature was allowed to return to 25°C?

(A) 1 atm
(B) 2 atm
(C) 3 atm
(D) 4 atm

20. A hydrocarbon was found to be 20% hydrogen by weight. If 1 mole of the hydrocarbon has a mass of 30 grams, what is its molecular formula?

(A) CH_2
(B) CH_3
(C) C_2H_4
(D) C_2H_6

21.
$$...CuFeS_2 + ...O_2 \rightarrow Cu_2S + ...FeO + ...SO_2$$

When the half-reaction above is balanced, what is the coefficient for O_2 if all the coefficients are reduced to the lowest whole number?

(A) 2
(B) 3
(C) 4
(D) 6

22. When chlorine gas is combined with fluorine gas, a compound is formed that is 38% chlorine and 62% fluorine. What is the empirical formula of the compound?

    (A) ClF
    (B) ClF_2
    (C) ClF_3
    (D) ClF_5

23. $$Cr_2O_7^{2-} + 6\ I^- + 14\ H^+ \rightarrow 2\ Cr^{3+} + 3\ I_2 + 7\ H_2O$$

    Which of the following statements about the reaction given above is NOT true?

    (A) The oxidation number of chromium changes from +6 to +3.
    (B) The oxidation number of iodine changes from −1 to 0.
    (C) The oxidation number of hydrogen changes from +1 to 0.
    (D) The oxidation number of oxygen remains the same.

24. $Al^{3+} + 3e^- \rightarrow Al(s)$ $E° = -1.66\ V$

    $Cr^{3+} + 3e^- \rightarrow Cr(s)$ $E° = -0.74\ V$

    The standard reduction potentials for two half-reactions are shown above. Which of the statements listed below will be true for the following reaction taking place under standard conditions?

    $$Al(s) + Cr^{3+} \rightarrow Al^{3+} + Cr(s)$$

    (A) $E° = 2.40\ V$, and the reaction is not spontaneous.
    (B) $E° = 0.92\ V$, and the reaction is spontaneous.
    (C) $E° = -0.92\ V$, and the reaction is not spontaneous.
    (D) $E° = -0.92\ V$, and the reaction is spontaneous.

25. In which of the following molecules does hydrogen have an oxidation state of −1?

    (A) H_2O
    (B) NH_3
    (C) CaH_2
    (D) CH_4

26. Oxygen takes the oxidation state −1 in hydrogen peroxide, H_2O_2. The equation for the decomposition of H_2O_2 is shown below.

    $$2\ H_2O_2 \rightarrow 2\ H_2O + O_2$$

    Which of the following statements about the reaction shown above is true?

    (A) Oxygen is reduced, and hydrogen is oxidized.
    (B) Oxygen is oxidized, and hydrogen is reduced.
    (C) Oxygen is both oxidized and reduced.
    (D) Hydrogen is both oxidized and reduced.

27. When solid iron is brought into contact with water and oxygen, it undergoes the following half-reaction:

$$Fe(s) \rightarrow Fe^{2+}(aq) + 2e^-$$

This half-reaction is instrumental in the corrosion of iron. When iron is coated with solid zinc, the half-reaction above is impeded, even if the zinc coating is incomplete. This is most likely because

(A) $Zn(s)$ is more easily reduced than $Fe(s)$.
(B) $Zn(s)$ is more easily oxidized than $Fe(s)$.
(C) $Zn^{2+}(aq)$ is more easily reduced than $Fe(s)$.
(D) $Zn^{2+}(aq)$ is more easily oxidized than $Fe(s)$.

28. $2 H_2O(l) + 2e^- \rightarrow H_2(g) + 2 OH^-(aq)$ $E° = -0.8$ V

$Na^+(aq) + e^- \rightarrow Na(s)$ $E° = -2.7$ V

$Cl_2(g) + 2e^- \rightarrow Cl^-(aq)$ $E° = +1.4$ V

Based on the reduction potentials given above, which of the following would be expected to occur when electrodes connected to the terminals of a 2.0 V battery are immersed in a solution of sodium chloride in water?

(A) Solid sodium will appear at the anode, and chlorine gas will appear at the cathode.
(B) Chlorine gas will appear at the anode, and solid sodium will appear at the cathode.
(C) Hydrogen gas will appear at the anode, and chlorine gas will appear at the cathode.
(D) Chlorine gas will appear at the anode, and hydrogen gas will appear at the cathode.

29. $Cu^{2+} + 2e^- \rightarrow Cu$ $E° = +0.3$ V

$Fe^{2+} + 2e^- \rightarrow Fe$ $E° = -0.4$ V

Based on the reduction potentials given above, what is the reaction potential for the following reaction?

$$Fe^{2+} + Cu \rightarrow Fe + Cu^{2+}$$

(A) -0.7 V
(B) -0.1 V
(C) $+0.1$ V
(D) $+0.7$ V

30. $Cu^{2+} + 2e^- \rightarrow Cu \qquad E° = +0.3$ V
$Zn^{2+} + 2e^- \rightarrow Zn \qquad E° = -0.8$ V
$Mn^{2+} + 2e^- \rightarrow Mn \qquad E° = -1.2$ V

Based on the reduction potentials given above, which of the following reactions will occur spontaneously?

(A) $Mn^{2+} + Cu \rightarrow Mn + Cu^{2+}$
(B) $Mn^{2+} + Zn \rightarrow Mn + Zn^{2+}$
(C) $Zn^{2+} + Cu \rightarrow Zn + Cu^{2+}$
(D) $Zn^{2+} + Mn \rightarrow Zn + Mn^{2+}$

31. $F_2 + 2e^- \rightarrow 2 F^- \qquad E° = 2.87$ V
$Ag^{2+} + e^- \rightarrow Ag^+ \qquad E° = 1.99$ V
$Au^{3+} + 3e^- \rightarrow 3 Au° \qquad E° = 1.50$ V
$Cl_2 + 2e^- \rightarrow 2 Cl^- \qquad E° = 1.36$ V
$Ag^+ + 3e^- \rightarrow 3 Ag° \qquad E° = 0.80$ V

In two separate experiments, pieces of metallic gold and metallic silver are reacted. In the first experiment both are reacted an inert solvent with Cl_2, while in the second they are reacted with F_2. Which of the following represents the metallic species present after the second reaction, which were not present after the first?

(A) Ag^+ only
(B) Ag^{2+} and Au^+
(C) Ag^+ and Ag^{2+}
(D) Ag^{2+} and Au^{3+}

32. Many metallic ions in the +1 oxidation state (generically stated, M^{+1}) are known to undergo *disproportionation* reactions in which two M^{+1} ions electrochemically react to form one M^{2+} and one $M°$. If this reaction for the generic metal M^+ is spontaneous at room temperature, which of the following 1 electron standard potentials *must* sum to a positive number?

(A) the reduction potential of M^{2+} and the reduction potential of M^+
(B the reduction potential of M^+ and the oxidation potential of M^+
(C) the oxidation potential of M^{2+} and the oxidation potential of M^+
(D) the oxidation potential of $M°$ and the reduction potential of M^{2+}

33. Molten $AlCl_3$ is electrolyzed with a constant current of 5.00 amperes over a period of 600.0 seconds. Which of the following expressions is equal to the maximum mass of Al(s) that plates out? (1 faraday = 96,500 coulombs)

(A) $\dfrac{(600)(5.00)}{(96,500)(3)(27.0)}$ grams

(B) $\dfrac{(600)(5.00)(3)(27.0)}{(96,500)}$ grams

(C) $\dfrac{(600)(5.00)(27.0)}{(96,500)(3)}$ grams

(D) $\dfrac{(96,500)(3)(27.0)}{(600)(5.00)}$ grams

Constructed Response Questions

1. A 10.0 gram sample containing calcium carbonate and an inert material was placed in excess hydrochloric acid. A reaction occurred producing calcium chloride, water, and carbon dioxide.

   (a) Write the balanced equation for the reaction.
   (b) When the reaction was complete, 900 milliliters of carbon dioxide gas were collected at 740 mmHg and 30°C. How many moles of calcium carbonate were consumed in the reaction?
   (c) If all of the calcium carbonate initially present in the sample was consumed in the reaction, what percent by mass of the sample was due to calcium carbonate?
   (d) If the inert material was silicon dioxide, what was the molar ratio of calcium carbonate to silicon dioxide in the original sample?

2. A gaseous hydrocarbon sample is completely burned in air, producing 1.80 liters of carbon dioxide at standard temperature and pressure and 2.16 grams of water.

   (a) What is the empirical formula for the hydrocarbon?
   (b) What was the mass of the hydrocarbon consumed?
   (c) The hydrocarbon was initially contained in a closed 1.00 liter vessel at a temperature of 32°C and a pressure of 760 millimeters of mercury. What is the molecular formula of the hydrocarbon?
   (d) Write the balanced equation for the combustion of the hydrocarbon.

3. The table below shows three common forms of copper ore.

Ore #	Empirical Formula	Percent by Weight		
		Copper	Sulfur	Iron
1	Cu_2S	?	?	0
2	?	34.6	34.9	30.5
3	?	55.6	28.1	16.3

   (a) What is the percent by weight of copper in Cu_2S?
   (b) What is the empirical formula of ore #2?
   (c) If a sample of ore #3 contains 11.0 grams of iron, how many grams of sulfur does it contain?
   (d) Cu can be extracted from Cu_2S by the following process:
   $$3\ Cu_2S + 3\ O_2 \rightarrow 3\ SO_2 + 6\ Cu$$
   If 3.84 grams of O_2 are consumed in the process, how many grams of Cu are produced?

4. $$2\ Mg(s) + 2\ CuSO_4(aq) + H_2O(l) \rightarrow 2\ MgSO_4(aq) + Cu_2O(s) + H_2(g)$$

   (a) If 1.46 grams of $Mg(s)$ are added to 500 milliliters of a 0.200-molar solution of $CuSO_4$, what is the maximum molar yield of $H_2(g)$?
   (b) When all of the limiting reagent has been consumed in (a), how many moles of the other reactant (not water) remain?
   (c) What is the mass of the Cu_2O produced in (a)?
   (d) What is the value of $[Mg^{2+}]$ in the solution at the end of the experiment? (Assume that the volume of the solution remains unchanged.)

5. A student performs an experiment in which a bar of unknown metal M is placed in a solution with the formula MNO_3. The metal is then hooked up to a copper bar in a solution of $CuSO_4$ as shown below. A salt bridge of KCl links the cell together.

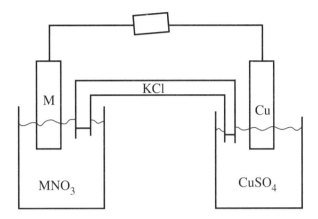

The potential for each cell was tested and found to have a value +0.74 V. Separately, when a bar of metal M is placed in the copper sulfate solution, solid copper starts to form on the bar. When a bar of copper is placed in the MNO_3 solution, no visible reaction occurs.

The following gives some reduction potentials for copper:

Half-reaction	$E°$
$Cu^{2+} + 2\,e^- \rightarrow Cu\ (s)$	0.34 V
$Cu^{2+} + e^- \rightarrow Cu^+$	0.15 V
$Cu^+ + e^- \rightarrow Cu\ (s)$	0.52 V

(a) Write the net ionic equation that takes place in the Cu/M cell.
(b) What is the standard reduction potential for metal M?
(c) Which metal acted as the anode and which as the cathode? Justify your answer.
(d) On the diagram of the cell, indicate which way the electrons are flowing in the wire. Additionally, indicate any ionic movement is occurring in the salt bridge.
(e) What would happen to the voltage of the reaction in the Cu/M cell if the concentration of the $CuSO_4$ increased while the concentration of the MNO_3 remained constant? Justify your answer.

6. Two electrodes are inserted into a solution of nickel (II) fluoride and a current of 2.20 A is run through them. A list of standard reduction potentials is as follows:

Half-reaction	$E°$
$O_2(g) + 4\,H^+ + 4e^- \rightarrow H_2O(l)$	1.23 V
$F_2(g) + 2e^- \rightarrow 2\,F^-$	2.87 V
$2\,H_2O(l) + 2e^- \rightarrow H_2\,(g) + 2\,OH^-$	–0.83 V
$Ni^{2+} + 2e^- \rightarrow Ni\,(s)$	–0.25 V

(a) Write the net ionic equation that takes place during this reaction.
(b) Qualitatively describe what an observer would see taking place at each electrode.
(c) Will the solution become acidic, basic, or remain neutral as the reaction progresses?
(d) How long would it take to create 1.2 g of Ni (s) at the cathode?

CHAPTER 5 ANSWERS AND EXPLANATIONS

Multiple-Choice Questions

1. **C** Z^{2+} was able to reduce to solid Z by taking electrons from metal X, so Z must have a higher reduction potential than X. Y^{2+} was unable to take electrons from metal X, and therefore Y must have a lower reduction potential than X.

2. **B** Carbonates are insoluble when paired with iron. Iron has a charge of 3+ and carbonate has a charge of 2–. To cancel out, both charges need to have a magnitude of 6, requiring two iron atoms and three carbonate ions.

3. **D** When those solutions mix, a precipitate of AgCl will form, removing those ions from the solution. The remaining ions do not react and remain the same as they were when they started.

4. **A** In an endothermic reaction, energy must be absorbed by the reactants for the reaction to occur. This is because the amount of energy necessary to break the bonds of the reactants is greater than the amount that is released by the formation of the bonds in the products.

5. **A** In (D), chromium is a pure element and has an oxidation number of 0. In (C), chromium's oxidation number is equal to its charge of +3, and in (B) it must balance the –2 on the oxygen, so it has a charge of +2. In (A), the total charge on the ion is –2, and each oxygen is – 2. Solving the following where X is the oxidation number on chromium: $X + -2(4) = -2$. So, the oxidation number is +6.

6. **D** An activated complex occurs when the bonds of the reactants have broken, but before the bonds in the products have formed. The level must always be above those of both the reactants and products. Option (A) is true for endothermic reactions, and option (B) is true for exothermic reactions, but neither option is correct for all reactions.

7. **D** The empirical formula of boron trifluoride is BF_3.

   Grams = (moles)(MW)
   Grams of boron = (1 mol)(10.8 g/mol) = 10.8 g
   Grams of fluorine = (3 mol)(19.0 g/mol) = 57.0 g
   So the mass ratio is about 57 to 11, which is about 5.3 to 1.

8. **A** The reaction will stop when no more precipitate can form, which will occur when either magnesium or phosphate ions run out. As there are no excess phosphate ions in solution after the reaction, the potassium phosphate must have run out first, leaving excess magnesium ions unreacted in solution.

9. **D** Moles = $\dfrac{\text{grams}}{\text{MW}}$

Moles of calcium hydroxide = $\dfrac{\left(148\ \text{g}\right)}{\left(74\ \text{g/mol}\right)}$ = 2 moles

Every mole of $Ca(OH)_2$ contains 2 moles of oxygen.

So there are (2)(2) = 4 moles of oxygen

Grams = (moles)(MW)

So grams of oxygen = (4 mol)(16 g/mol) = 64 grams

10. **A** Assume that we have 100 grams of the compound. That means that we have 31 grams of oxygen and 69 grams of chlorine.

Moles = $\dfrac{\text{grams}}{\text{MW}}$

Moles of oxygen = $\dfrac{\left(31\text{g}\right)}{\left(16\ \text{g/mol}\right)}$ = slightly less than 2 mol

Moles of chlorine = $\dfrac{\left(69\ \text{g}\right)}{\left(35.5\ \text{g/mol}\right)}$ = slightly less than 2 mol

So the ratio of chlorine to oxygen is 1 to 1, and the empirical formula is ClO^-.

11. **C** $12.25\ \text{g KClO}_3 \times \dfrac{1\ \text{mol KClO}_3}{122.25\ \text{g KClO}_3} = 0.1000\ \text{mol KClO}_3$

$0.1000\ \text{mol KClO}_3 \times \dfrac{3\ \text{mol O}_2}{2\ \text{mol KClO}_3} = 0.1500\ \text{mol O}_2$

$0.1500\ \text{mol O}_2 \times \dfrac{32.00\ \text{g O}_2}{1\ \text{mol O}_2} = 4.80\ \text{g O}_2$

Note that even without a calculator, you are expected to be able to do simple calculations such as the ones above.

12. **B** $0.1500\ \text{mol O}_2 \times \dfrac{22.4\ \text{L}}{1\ \text{mol O}_2} = 3.36\ \text{L O}_2$

13. **A** Temperature is a measure of the speed of the molecules, and pressure is a measure of how much momentum the gas particles are hitting the wall with. If the molecules are moving twice as fast, they will hit the wall twice as often, thus doubling the pressure.

14. **D** A catalyst does not change the amount of bond energy present in either the reactants or the products, and catalysts always speed up the overall reaction. Many decomposition reactions occur more quickly with a catalyst, but even without one present the reaction can progress, albeit at a slower rate.

15. **C** The molecular weight of $CuSO_4$ is 160 g/mol, so we have only 1 mole of the hydrate. The lost mass was due to water, so 1 mole of the hydrate must have contained 90 grams of H_2O.

$$Moles = \frac{grams}{MW}$$

$$Moles\ of\ water = \frac{(90\ g)}{(18\ g/mol)} = 5\ moles$$

So if 1 mole of hydrate contains 5 moles of H_2O, then the formula for the hydrate must be $CuSO_4 \cdot 5\ H_2O$.

16. **B** You might be able to do this one in your head just from knowing that sulfur's molecular weight is twice as large as oxygen's. If not, let's say you have 100 grams of the compound. So you have 50 grams of sulfur and 50 grams of oxygen.

$$Moles = \frac{grams}{MW}$$

$$Moles\ of\ sulfur = \frac{(50\ g)}{(32\ g/mol)} = a\ little\ more\ than\ 1.5$$

$$Moles\ of\ oxygen = \frac{(50\ g)}{(16\ g/mol)} = a\ little\ more\ than\ 3$$

The molar ratio of O to S is 2 to 1, so the empirical formula must be SO_2.

17. **B** Use trial and error or backsolve.

Start with (C). If there are 3 OH^-, there can't be a whole number coefficient for H_2O, so (C) is wrong. You should also be able to see that the answer can't be an odd number, so (A) is are also wrong.

Try (D). If there are 4 OH^-, then there are 2 H_2O.

That leaves 2 more Os on the product side, so there must be 2 CNO⁻.

If there are 2 CNO⁻ then there are 2 CN⁻.

These are all whole numbers, but they are not the lowest whole numbers, so (D) is wrong.

If we divide all the coefficients by 2, we get the lowest whole number coefficients. That leaves us with 2 OH⁻, which is choice (B).

By the way, N^{5-} (in CN^-) is oxidized to N^{3-} (in CNO^-), so there are 2 e^-.

18. **C** $Moles = \dfrac{grams}{MW}$

Moles of $CaCO_3 = \dfrac{(150 \text{ g})}{(100 \text{ g/mol})} = 1.5$ moles

From the balanced equation, for every mole of $CaCO_3$ consumed, one mole of CO_2 is produced. So 1.5 moles of CO_2 are produced.

At STP, volume of gas = (moles)(22.4 L)

So volume of CO_2 = (1.5)(22.4) = 34 L

19. **B** All of the reactants are consumed in the reaction and the temperature doesn't change, so the pressure will change only if the number of moles of gas changes over the course of the reaction. The number of moles of gas (3 moles) doesn't change in the balanced equation, so the pressure will remain the same (2 atm) at the end of the reaction as at the beginning.

20. **D** Let's say we have 100 grams of the compound.

$Moles = \dfrac{grams}{MW}$

So moles of carbon = $\dfrac{(80 \text{ g})}{(12 \text{ g/mol})} = 6.7$ moles

and moles of hydrogen = $\dfrac{(20 \text{ g})}{(1 \text{ g/mol})} = 20$ moles

According to our rough calculation, there are about three times as many moles of hydrogen in the compound as there are moles of carbon, so the empirical formula is CH_3.

The molar mass for the empirical formula is 15 g/mol, so we need to double the moles of each element to get a compound with a molar mass of 30 g/mol. That makes the molecular formula of the compound C_2H_6.

21.　C　Backsolving doesn't work so well in this case because there are two different compounds that contain oxygen on the right side of the equation, which makes the process kind of confusing. Instead, let's just try plugging in values for the most complicated compound in the equation, $CuFeS_2$.

What if there's 1 $CuFeS_2$? That's impossible because there are 2 Cu's on the right.

What if there are 2 $CuFeS_2$'s? Then the right side has 1 Cu_2S to balance the Cu and 2 FeO to balance the Fe. The right side must also have 3 SO_2 to balance the S.

Now there are 8 O's on the right, so there must be 4 O_2's on the left, and the equation is balanced.

$$2\ CuFeS_2 + 4\ O_2 \rightarrow 1\ Cu_2S + 2\ FeO + 3\ SO_2$$

22.　C　Let's say we have 100 grams of the compound.

$$Moles = \frac{grams}{MW}$$

$$Moles\ of\ chlorine = \frac{(38\ g)}{(35.5\ g/mol)} = about\ 1\ mole$$

$$Moles\ of\ fluorine = \frac{(62\ g)}{(19\ g/mol)} = about\ 3\ moles$$

According to our rough calculation, there are about three times as many moles of fluorine in the compound as there are moles of chlorine, so the empirical formula is ClF_3.

23.　C　The oxidation numbers of the reactants are Cr^{6+}, O^{2-}, I^-, and H^+, and the oxidation numbers of the products are Cr^{3+}, O^{2-}, I^0, and H^+.

Chromium gains electrons and is reduced, and iodine loses electrons and is oxidized; the oxidation states of oxygen and hydrogen are not changed.

24.　B　$E°$ for a redox reaction is given by the expression below.

$$E° = E°_{ox} + E°_{red}$$

Al(*s*) loses 3 electrons in the reaction, so it is oxidized (LEO), and we use the voltage given for the reduction half-reaction, but we change the sign. So $E°_{ox} = 1.66$ V.

Cr^{3+} gains 3 electrons in the reaction, so it is reduced (GER), and we use the voltage given for the reduction half-reaction. So $E°_{red} = -0.74$ V.

So, $E° = 1.66$ V + $(-0.74$ V$) = 0.92$ V.

From the relationship, $\Delta G° = -nFE°$, we know that if $E°$ is positive, then $\Delta G°$ is negative, and if $\Delta G°$ is negative, then the reaction is spontaneous under standard conditions.

25. **C** In a molecule, the more electronegative element takes the negative oxidation state. Hydrogen is more electronegative than calcium, so the oxidation state for Ca is +2 and the oxidation state for H is –1.

In choices (A), (B), and (D), hydrogen is the less electronegative element, and it takes the oxidation state of +1.

26. **C** The oxidation state of hydrogen remains +1 throughout the reaction.

We have O^- at the start.

When water forms, oxygen has gained an electron and been reduced to O^{2-} (GER). When O_2 forms, oxygen has been oxidized to O^0 through the loss of an electron (LEO). As you can see, in this process, oxygen is both oxidized and reduced.

27. **B** The half-reaction given above is for the oxidation of Fe (LEO). If the presence of Zn impedes this process it is because Zn is oxidized instead of Fe. Iron nails are often coated with zinc to help keep them from rusting.

28. **D** A voltage of 2.0 V is not enough to cause the reduction of Na^+ (more than 2.7 V is needed), so no solid sodium will form. A voltage of 2.0 V will cause the reduction of H_2O, which forms hydrogen gas, and the oxidation of Cl^-, which forms chlorine gas.

From AN OX and RED CAT, we know that hydrogen gas is formed at the cathode and that chlorine gas is formed at the anode.

29. **A** We add the reduction potential for Fe^{2+} (–0.4 V) to the oxidation potential for Cu (–0.3 V, the reverse of the reduction potential) to get –0.7 V.

30. **D** To get the reaction potential, we add the reduction potential for the reduction half-reaction to the oxidation potential for the oxidation half–reaction.

Here are the reaction potentials for choices (A)–(D).

(A) $(-1.2 \text{ V}) + (-0.3 \text{ V}) = -1.5 \text{ V}$

(B) $(-1.2 \text{ V}) + (+0.8 \text{ V}) = -0.4 \text{ V}$

(C) $(-0.8 \text{ V}) + (-0.3 \text{ V}) = -1.1 \text{ V}$

(D) $(-0.8 \text{ V}) + (+1.2 \text{ V}) = +0.4 \text{ V}$

Choice (D) is the only reaction with a positive voltage, which means that it is the only reaction listed that occurs spontaneously.

31. **D** In order for the metallic species to be oxidized the reduction potential of the oxidant (F_2 or Cl_2) must be greater than the reduction potential of the metallic species. Cl_2 is a strong enough oxidant to take Ag^o to Ag^+, but not strong enough to take Ag^+ to Ag^{2+} or Au^o to Au^{3+}. On the other hand, F_2 *is* a strong enough oxidant, as it has the highest reduction potential listed, to do all of the above electrochemical reactions. Therefore Au^o will be oxidized to Au^{3+} and Ag^o will be oxidized all the way to Ag^{2+}.

32. **B** The spontaneity of the disproportionation reaction means that the standard potentials of the following two half reactions must sum to a positive number:

$M^+ + 1e^- \rightarrow M^o$

$M^+ \rightarrow 1e^- + M^{2+}$

The two potentials here are the reduction potential and oxidation potential of M^+. Without knowing the relative values and signs of the two individual reactions we cannot say for sure any of the rest of the listed pairs of potentials would sum to a positive.

33. **C** First let's find out how many electrons are provided by the current.

Coulombs = (seconds)(amperes) = $(600)(5.00)$

Moles of electrons = $\dfrac{(\text{coulombs})}{(96,500)} = \dfrac{(600)(5.00)}{(96,500)}$ moles

In $AlCl_3$, we have Al^{3+}, so the half-reaction for the plating of aluminum is

$Al^{3+} + 3e^- \rightarrow Al(s)$

So we get $\dfrac{1}{3}$ as many moles of Al(s) as we have moles of electrons.

Moles of Al(s) = (moles of electrons)$\left(\dfrac{1}{3}\right)$ = $\dfrac{(600)(5.00)}{(96,500)(3)}$ moles

Now, grams = (moles)(MW)

Grams of Al(s) = $\dfrac{(600)(5.00)}{(96,500)(3)}(27.0)$ = $\dfrac{(600)(5.00)(27.0)}{(96,500)(3)}$ grams

Constructed Response Questions

1. (a) $CaCO_3 + 2\,HCl \rightarrow CaCl_2 + H_2O + CO_2$

   (b) Use the ideal gas equation to find the number of moles of CO_2 produced. Remember to convert to the proper units.

   $$n = \dfrac{PV}{RT} = \dfrac{\left(\dfrac{740}{760}\ \text{atm}\right)(0.900\ \text{L})}{(0.0821)(303\ \text{K})} = 0.035\ \text{moles}$$

   From the balanced equation, for every mole of CO_2 produced, 1 mole of $CaCO_3$ was consumed. So 0.035 moles of $CaCO_3$ were consumed.

   (c) We know the number of moles of $CaCO_3$, so we can find the mass.

   Grams = (moles)(MW)

   Grams of $CaCO_3$ = (0.035 mol)(100 g/mol) = 3.50 grams

   Percent by mass = $\dfrac{\text{mass of } CaCO_3}{\text{mass of sample}} \times 100 = \dfrac{(3.50\ \text{g})}{(10.0\ \text{g})} \times 100 = 35\%$

   (d) Mass of SiO_2 = 10.0 g – 3.5 g = 6.5 g

   Moles = $\dfrac{\text{grams}}{\text{MW}}$

   Moles of SiO_2 = $\dfrac{(6.5\ \text{g})}{(60\ \text{g/mol})} = 0.11\ \text{mol}$

   Molar ratio = $\dfrac{\text{moles of } CaCO_3}{\text{moles of } SiO_2} = \dfrac{(0.035)}{(0.11)} = 0.32$

2. (a) All of the hydrogen in the water and all of the carbon in the carbon dioxide must have come from the hydrocarbon.

$$\text{Moles of } H_2O = \frac{(2.16 \text{ g})}{(18.0 \text{ g/mol})} = 0.120 \text{ moles}$$

Every mole of water contains 2 moles of hydrogen, so there are 0.240 moles of hydrogen.

$$\text{Moles of } CO_2 = \frac{(1.80 \text{ L})}{(22.4 \text{ L/mol})} = 0.080 \text{ moles}$$

Every mole of CO_2 contains 1 mole of carbon, so there are 0.080 moles of carbon.

There are three times as many moles of hydrogen as there are moles of carbon, so the empirical formula of the hydrocarbon is CH_3.

(b) In (a), we found the number of moles of hydrogen and carbon consumed, so we can find the mass of the hydrocarbon.

Grams = (moles)(MW)

Grams of H = (0.240 mol)(1.01 g/mol) = 0.242 g

Grams of C = (0.080 mol)(12.01 g/mol) = 0.961 g

Grams of hydrocarbon = (0.242) + (0.961) = 1.203 g

(c) First let's find the number of moles of hydrocarbon from the ideal gas law. Don't forget to convert to the appropriate units (760 mmHg = 1 atm, 32°C = 305 K).

$$n = \frac{PV}{RT} = \frac{(1.00 \text{ atm})(1.00 \text{ L})}{(0.0821)(305 \text{ K})} = 0.040 \text{ moles}$$

Now we can use the mass we found in (b) to find the molecular weight of the hydrocarbon.

$$MW = \frac{\text{grams}}{\text{moles}} = \frac{(1.203 \text{ g})}{(0.040 \text{ mol})} = 30.1 \text{ g/mole}$$

CH_3 would have a molecular weight of 15, so we can just double the empirical formula to get the molecular formula, which is C_2H_6.

(d) $2 C_2H_6 + 7 O_2 \rightarrow 4 CO_2 + 6 H_2O$

3. (a) First find the molecular weight of Cu_2S.

MW of Cu_2S = (2)(63.6) + (1)(32.1) = 159.3% by mass of

$$Cu = \frac{\text{mass of } Cu}{\text{mass of } Cu_2S} \times 100$$

$$= \frac{(2)(63.6)}{(159.3)} \times 100 = 79.8\%$$

(b) Assume that we have 100 grams of ore #2. So we have 34.6 g of Cu, 30.5 g of Fe, and 34.9 g of S. To get the empirical formula, we need to find the number of moles of each element.

$$\text{Moles} = \frac{\text{grams}}{\text{MW}}$$

$$\text{Moles of } Cu = \frac{(34.6 \text{ g})}{(63.6 \text{ g/mol})} = 0.544 \text{ moles of Cu}$$

$$\text{Moles of } Fe = \frac{(30.5 \text{ g})}{(55.9 \text{ g/mol})} = 0.546 \text{ moles of Fe}$$

$$\text{Moles of } S = \frac{(34.9 \text{ g})}{(32.1 \text{ g/mol})} = 1.09 \text{ moles of S}$$

So the molar ratio of Cu:Fe:S is 1:1:2, and the empirical formula for

ore #2 is $CuFeS_2$.

(c) You can use the ratio of the percents by weight.

$$\text{Mass of } S = \frac{\text{\% by mass of S}}{\text{\% by mass of Fe}} \times (\text{mass of Fe}) = \frac{(28.1\,\%)}{(16.3\,\%)}(11.0 \text{ g}) = 19.0 \text{ g}$$

(d) First find the moles of O_2 consumed.

$$\text{Moles} = \frac{\text{grams}}{\text{MW}}$$

$$\text{Moles of } O_2 = \frac{(3.84 \text{ g})}{(32.0 \text{ g/mol})} = 0.120 \text{ moles}$$

From the balanced equation, for every 3 moles of O_2 consumed, 6 moles of Cu are produced, so the number of moles of Cu produced will be twice the number of moles of O_2 consumed. So 0.240 moles of Cu are produced.

Grams = (moles)(MW)

Grams of Cu = (0.240 mol)(63.6 g/mol) = 15.3 grams

4. (a) We need to find the limiting reagent. There's plenty of water, so it must be one of the other two reactants.

$$\text{Moles} = \frac{\text{grams}}{\text{MW}}$$

$$\text{Moles of Mg} = \frac{(1.46\ \text{g})}{(24.3\ \text{g/mol})} = 0.060\ \text{moles}$$

$$\text{Moles} = (\text{molarity})(\text{volume})$$

$$\text{Moles of CuSO}_4 = (0.200\ \text{M})(0.500\ \text{L}) = 0.100\ \text{moles}$$

From the balanced equation, Mg and $CuSO_4$ are consumed in a 1:1 ratio, so we'll run out of Mg first. Mg is the limiting reagent, and we'll use it to find the yield of H_2.

From the balanced equation, 1 mole of H_2 is produced for every 2 moles of Mg consumed, so the number of moles of H_2 produced will be half the number of moles of Mg consumed.

$$\text{Moles of H}_2 = \frac{1}{2}\ (0.060\ \text{mol}) = 0.030\ \text{moles}$$

(b) Mg is the limiting reagent, so some $CuSO_4$ will remain. From the balanced equation, Mg and $CuSO_4$ are consumed in a 1:1 ratio, so when 0.060 moles of Mg are consumed, 0.060 moles of $CuSO_4$ are also consumed.

$$\text{Moles of CuSO}_4 \text{ remaining} = (0.100\ \text{mol}) - (0.060\ \text{mol}) = 0.040\ \text{moles}$$

(c) From the balanced equation, 1 mole of Cu_2O is produced for every 2 moles of Mg consumed, so the number of moles of Cu_2O produced will be half the number of moles of Mg consumed.

$$\text{Moles of Cu}_2\text{O} = \frac{1}{2}\ (0.060\ \text{mol}) = 0.030\ \text{moles}$$

$$\text{Grams} = (\text{moles})(\text{MW})$$

$$\text{Grams of Cu}_2\text{O} = (0.030\ \text{mol})(143\ \text{g/mol}) = 4.29\ \text{grams}$$

(d) All of the Mg consumed ends up as Mg^{2+} ions in the solution.

$$\text{Molarity} = \frac{\text{moles}}{\text{liters}}$$

$$[Mg^{2+}] = \frac{(0.060\ \text{mol})}{(0.500\ \text{L})} = 0.120\ \text{M}$$

5. (a) When metal M was placed in the copper solution, a reaction occurred. Therefore, the copper must have the higher reduction potential and thus is reduced in the Cu/M cell. The half reactions are:

Reduction: $Cu^{2+} + 2e^- \rightarrow Cu\ (s)$
Oxidation: $M(s) \rightarrow M^+ + e^-$

The oxidation half-reaction must be multiplied by two to balance the electrons before combining the reactions to yield:

$Cu^{2+} + 2\ M(s) \rightarrow Cu\ (s) + 2\ M^+$

(b) $E_{red} + E_{ox} = +0.74\ V$

The reduction potential for $Cu^{2+} + 2e^- \rightarrow Cu\ (s)$ is known:

$0.34\ V + E_{ox} = +0.74\ V$

$E_{ox} = +0.40\ V$

The reduction potential for metal M is the opposite of its oxidation potential.

$E_{red} = -0.40\ V$

(c) Reduction occurs at the cathode, so copper is the cathode. Oxidation occurs at the anode, so M is the anode.

(d)

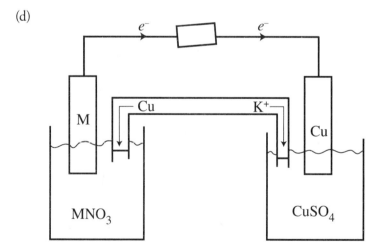

Electrons should be flowing from metal M to the copper bar. In the salt bridge, the K^+ ions will flow towards the copper solution to replace the Cu^{2+} being reduced into $Cu(s)$. The Cl^- will flow towards the solution for metal M in order to balance out the charge of the extra M^+ ions being created via the oxidation of $M(s)$.

(e) If the concentration of the Cu^{2+} increases, it will cause a shift to the right in accordance with Le Châtelier's principle. This will increase the cell's overall voltage.

6. (a) The two potential reduction reactions are the bottom two, as their reactants (H_2O and Ni^{2+}) are actually present at the start of the reaction. Of the two, the nickel reduction is more positive, and thus will take place. The top two reactions must be flipped to have the reactants (H_2O and F^-) on the reactants side, which also flips the sign. the sign the water oxidation will have a more positive value and thus will occur.

Reduction: $Ni^{2+} + 2e^- \rightarrow Ni\ (s)$
Oxidation: $H_2O(l) \rightarrow O_2(g) + 4\ H^+ + 4e^-$

After multiplying the reduction half reaction by two to balance the electrons and combining both half reactions, the net cell reaction is:

$2\ Ni^{2+} + H_2O\ (l) \rightarrow 2\ Ni\ (s) + O_2\ (g) + 4\ H^+$

(b) At the cathode, solid nickel would begin to plate out of solution. At the anode, oxygen gas would form and bubble up to the surface.

(c) The solution will become more acidic due to the creation of hydrogen ions at the anode

(d) $1.20 \text{ g Ni} \times \dfrac{1 \text{ mol Ni}}{58.69 \text{ g Ni}} \times \dfrac{2 \text{ mol } e^-}{1 \text{ mol Ni}} \times \dfrac{96500 \text{ C}}{1 \text{ mol } e^-} \times \dfrac{1.0 \text{ s}}{2.20 \text{ C}} = 1790 \text{ seconds}$

Chapter 6
Big Idea #4: Chemical Reactions and their Rates

Rates of chemical reactions are determined by details of the molecular collisions.

RATE LAW USING INITIAL CONCENTRATIONS

The rate law for a reaction describes the dependence of the initial rate of a reaction on the concentrations of its reactants. It includes the Arrhenius constant, k, which takes into account the activation energy for the reaction and the temperature at which the reaction occurs. The rate of a reaction is described in terms of the rate of appearance of a product or the rate of disappearance of a reactant. The rate law for a reaction cannot be determined from a balanced equation; it must be determined from experimental data, which is presented on the test in table form.

Here's how it's done

The data below were collected for the following hypothetical reaction:

$$A + 2 B + C \rightarrow D$$

Experiment	Initial Concentration of Reactants (M)			Initial Rate of Formation of D (M/sec)
	[A]	[B]	[C]	
1	0.10	0.10	0.10	0.01
2	0.10	0.10	0.20	0.01
3	0.10	0.20	0.10	0.02
4	0.20	0.20	0.10	0.08

The rate law always takes the following form, using the concentrations of the reactants:

$$\text{Rate} = k[A]^x[B]^y[C]^z$$

The greater the value of a reactant's exponent, the more a change in the concentration of that reactant will affect the rate of the reaction. To find the values for the exponents x, y, and z, we need to examine how changes in the individual reactants affect the rate. The easiest way to find the exponents is to see what happens to the rate when the concentration of an individual reactant is doubled.

Let's look at [A]

From experiment 3 to experiment 4, [A] doubles while the other reactant concentrations remain constant. For this reason, it is useful to use the rate values from these two experiments to calculate x (the order of the reaction with respect to reactant A).

As you can see from the table, the rate quadruples from experiment 3 to experiment 4, going from 0.02 M/sec to 0.08 M/sec.

We need to find a value for the exponent x that relates the doubling of the concentration to the quadrupling of the rate. The value of x can be calculated as follows:

$$(2)^x = 4, \text{ so } x = 2$$

Because the value of x is 2, the reaction is said to be second order with respect to A.

$$\text{Rate} = k[A]^2[B]^y[C]^z$$

Let's look at [B]

From experiment 1 to experiment 3, [B] doubles while the other reactant concentrations remain constant. For this reason it is useful to use the rate values from these two experiments to calculate y (the order of the reaction with respect to reactant B).

As you can see from the table, the rate doubles from experiment 1 to experiment 3, going from 0.01 M/sec to 0.02 M/sec.

We need to find a value for the exponent y that relates the doubling of the concentration to the doubling of the rate. The value of y can be calculated as follows:

$$(2)^y = 2, \text{ so } y = 1$$

Because the value of y is 1, the reaction is said to be first order with respect to B.

$$\text{Rate} = k[A]^2[B][C]^z$$

Let's look at [C]

From experiment 1 to experiment 2, [C] doubles while the other reactant concentrations remain constant.

The rate remains the same at 0.01 M.

The rate change is $(2)^z = 1$, so z = 0.

Because the value of z is 0, the reaction is said to be zero order with respect to C.

$$\text{Rate} = k[A]^2[B]$$

Because the sum of the exponents is 3, the reaction is said to be third order overall.

Once the rate law has been determined, the value of the rate constant can be calculated using any of the lines of data on the table. The units of the rate constant are dependent on the order of the reaction, so it's important to carry along units throughout all rate constant calculations.

Let's use experiment 3.

$$k = \frac{\text{Rate}}{[A]^2[B]} = \frac{\left(0.02\,M\,/\,\sec\right)}{\left(0.10\,M\right)^2\left(0.20\,M\right)} = 10\left(\frac{(M)}{(M)^3(\sec)}\right) = 10\,M^{-2}\text{-sec}^{-1}$$

You should note that we can tell from the coefficients in the original balanced equation that the rate of appearance of D is equal to the rate of disappearance of A and C because the coefficients of all three are the same. The coefficient of D is half as large as the coefficient of B, however, so the rate at which D appears is half the rate at which B disappears.

RATE LAW USING CONCENTRATION AND TIME

It's also useful to have rate laws that relate the rate constant k to the way that concentrations change over time. The rate laws will be different depending on whether the reaction is first, second, or zero order, but each rate law can be expressed as a graph that relates the rate constant, the concentration of a reactant, to the elapsed time.

First-Order Rate Laws

The rate of a first-order reaction depends on the concentration of a single reactant raised to the first power.

$$\text{Rate} = k[A]$$

As the concentration of reactant A is depleted over time, the rate of reaction will decrease with a characteristic half-life. This is the same curve you've seen used for nuclear decay, which is also a first-order process.

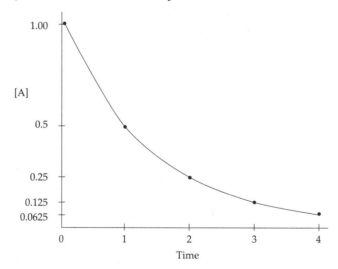

The rate law for a first-order reaction uses natural logarithms.

The use of natural logarithms in the rate law creates a linear graph comparing concentration and time. The slope of the line is given by $-k$ and the y-intercept is given by $\ln[A]_o$.

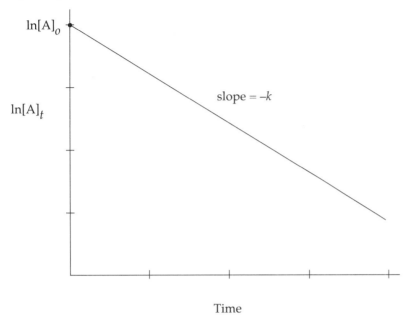

Using slope intercept form, we can interpret this graph to come up with a useful equation.

$$y = mx + b$$

$y = \ln[A]_t$
$m = -k$
$x = \text{time}$
$b = \ln[A]_0$

After substitution and rearrangement, the slope intercept equation becomes:

First-Order Rate Law
$$\ln[A]_t = -kt + \ln[A]_o$$

$[A]_t =$ concentration of reactant A at time t
$[A]_o =$ initial concentration of reactant A
$k =$ rate constant
$t =$ time elapsed

Half-Life

The half-life of a reactant in a chemical reaction is the time it takes for half of the substance to react. Most half-life-problems can be solved by using a simple chart:

(Don't forget that the chart should start with the time at zero.)

Time	Sample
0	100%
1 half-life	50%
2 half-lives	25%
3 half-lives	12.5%

So a sample with a mass of 120 grams and a half-life of 3 years will decay as follows:

Time	Sample
0	120 g
3	60 g
6	30 g
9	15 g

The fact that you're not allowed to use a calculator for Section I and that you're not given any half-life formulas for Section II means that you should be able to solve any half-life problem that comes up by using the chart and POE.

Additionally, the half-life of a first order reactant can be calculated from the graphical interpretation of the reaction using the following formula:

$$\text{Half-life} = \frac{\ln 2}{k} = \frac{0.693}{k}$$

Let's try an example based on the data below.

[A] (M)	Time (min)
2.0	0
1.6	5
1.2	10

(a)   Let's find the value of k. We'll use the first two lines of the table.
$\ln[A]_t - \ln[A]_o = -kt$
$\ln(1.6) - \ln(2.0) = -k(5 \text{ min})$
$-0.22 = -(5 \text{ min})k$
$k = 0.045 \text{ min}^{-1}$

(b) Now let's use k to find [A] when 20 minutes have elapsed.

$\ln[A]_t - \ln[A]_o = -kt$

$\ln[A]_t - \ln(2.0) = -(0.045 \text{ min}^{-1})(20 \text{ min})$

$\ln[A]_t = -0.21$

$[A]_t = e^{-0.21} = 0.81 \ M$

(c) Now let's find the half-life of the reaction.

We can look at the answer (b) to see that the concentration dropped by half (1.6 M to 0.8 M) from 5 minutes to 20 minutes. That makes the half-life about 15 minutes. We can confirm this using the half-life equation.

$$\text{Half-life} = \frac{0.693}{k} = \frac{0.693}{0.045 \text{ min}^{-1}} = 15.4 \text{ minutes}$$

Another use of half-life is to examine the rate of decay of a radioactive substance. A radioactive substance is one which will slowly decay into a more stable form as time goes on.

Second-Order Rate Laws

The rate of a second-order reaction depends on the concentration of a single reactant raised to the second power.

$$\text{Rate} = k[A]^2$$

The rate law for a second-order reaction uses the inverses of the concentrations.

Second-Order Rate Law

$$\frac{1}{[A]_t} = -kt + \frac{1}{[A]_o}$$

$[A]_t$ = concentration of reactant A at time t

$[A]_o$ = initial concentration of reactant A

k = rate constant

t = time elapsed

The use of inverses in the rate law creates a linear graph comparing concentration and time.

Notice that the line moves upward as the concentration decreases. The slope of the line is given by k and the y-intercept is given by $\frac{1}{[A]_o}$.

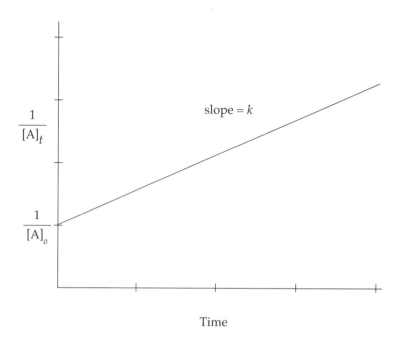

Zero-Order Rate Laws

The rate of a zero-order reaction does not depend on the concentration of reactants at all, so the rate of a zero-order reaction will always be the same at a given temperature.

$$Rate = k$$

The graph of the change in concentration of a reactant of a zero-order reaction versus time will be a straight line with a slope equal to $-k$.

COLLISION THEORY

According to collision theory, chemical reactions occur because reactants are constantly moving around and colliding with one another.

When reactants collide with sufficient energy (**activation energy, E_a**), a reaction occurs. These collisions are referred to as effective collisions, because they lead to a chemical reaction. Ineffective collisions do not produce a chemical reaction. At any given time during a reaction, a certain fraction of the reactant molecules will collide with sufficient energy to cause a reaction between them.

Reaction rate increases with increasing concentration of reactants because if there are more reactant molecules moving around in a given volume, then more collisions will occur.

Reaction rate increases with increasing temperature because increasing temperature means that the molecules are moving faster, which means that the molecules have greater average kinetic energy. The higher the temperature, the greater the number of reactant molecules colliding with each other with enough energy (E_a) to cause a reaction.

The Boltzmann distribution diagram below is often used to show that increasing temperature increases the fraction of reactant molecules above the activation energy.

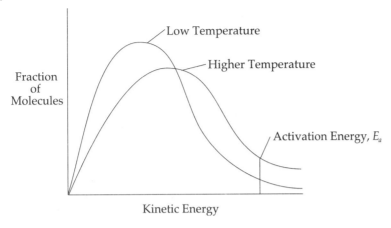

BEER'S LAW

To measure the concentration of a solution over time, a device called a spectrophotometer can be used in some situations. A spectrophotometer measures the amount of light at a given wavelength that is absorbed by a solution. If a solution changes color as the reaction progresses, the amount of light that is absorbed will change. Absorbance can be calculated using Beer's Law:

Beer's Law
$$A = abc$$

A = absorbance
a = molar absorptivity, a constant that depends on the solution
b = path length, the distance the light is traveling through the solution
c = concentration of the solution

As molar absorptivity and path length are constants when using a spectrophotometer, Beer's Law is often interpreted as a direct relationship between absorbance and the concentration of the solution. Beer's Law is most effective with solutions that visibly change color over the course of a reaction, but if a spectrophotometer that emits light in the ultraviolet region is used Beer's Law can be used to determine the concentrations of reactants in solutions that are invisible to the human eye.

REACTION MECHANISMS

Many chemical reactions are not one-step processes. Rather, the balanced equation is the sum of a series of individual steps, called elementary steps.

For instance, the hypothetical reaction

$$2\,A + 2\,B \rightarrow C + D \qquad\qquad \text{Rate} = k[A]^2[B]$$

could take place by the following three-step mechanism:

I.	$A + A \rightleftharpoons X$	(fast)
II.	$X + B \rightarrow C + Y$	(slow)
III.	$Y + B \rightarrow D$	(fast)

Species X and Y are called **intermediates** because they appear in the mechanism, but they cancel out of the balanced equation. The steps of a reaction mechanism must add up to equal the balanced equation, with all intermediates cancelling out.

Let's show that the mechanism above is consistent with the balanced equation by adding up all the steps.

I. $A + A \rightleftharpoons X$

II. $X + B \rightarrow C + Y$

III. $Y + B \rightarrow D$

————————————————————————

$A + A + X + B + Y + B \rightarrow X + C + Y + D$

Cancel species that appear on both sides.

$$2\,A + 2\,B \rightarrow C + D$$

By adding up all the steps, we get the balanced equation for the overall reaction, so this mechanism is consistent with the balanced equation.

As in any process where many steps are involved, the speed of the whole process can't be faster than the speed of the slowest step in the process, so the slowest step of a reaction is called the **rate-determining step.** Because the slowest step is the most important step in determining the rate of a reaction, the slowest step and the steps leading up to it are used to see if the mechanism is consistent with the rate law for the overall reaction.

The rate for an elementary step can be determined using by taking the concentration of the reactants in that step and raising them to the power of any coefficient attached to that reactant. So, for the current reactions:

I. $A + A \leftrightarrow X$ (so $2A \leftrightarrow X$)   Rate = $k[A]^2$ (fast)

II. $X + B \rightarrow Y$ Rate = $k[X][B]$ (slow)

III. $Y + B \rightarrow D$ Rate = $k[Y][B]$ (fast)

The rate law for the entire reaction is equal to that of the slowest elementary step, which is step II. However, step II has an intermediate (X) present in it. Looking at step 1, we can also see that $[X]$ is equivalent to $[A]^2$ (as the sides are in equilibrium). If $[A]^2$ is substituted in for $[X]$, the rate law for step II because rate = $k[A]^2[B]$, which is the overall rate law for this reaction.

It is important to emphasize that you can only use the coefficient method to determine the rate law of elementary steps. You CANNOT use it to determine the rate law of an overall reaction. Overall rate laws can only be determined via experimental data.

CATALYSTS

As we mentioned earlier in the book, a catalyst increases the rate of a chemical reaction without being consumed in the process; catalysts do not appear in the balanced equation. In some cases, a catalyst is a necessary part of a reaction because in its absence, the reaction would proceed at too slow a rate to be at all useful.

When looking at elementary steps, a catalyst is present both before and after the overall reaction. Catalysts cancel out of the overall reaction but are present in elementary steps.

Example: $A + B \rightarrow C$

I. $A + X \rightarrow Y$

II. $B + Y \rightarrow C + X$

In the above example, X is a catalyst and Y is a reaction intermediate.

When a catalyst is introduced to a reaction, the ensuing reaction is said to undergo catalysis. There are many types of catalysis. One of the most common is surface catalysis, in which a reaction intermediate is formed as in the example above. Another is enzyme catalysis, in which the catalyst binds to the reactants in a way to reduce the overall activation energy of the reaction. Enzymes are very common in biological applications. Finally, in acid-base catalysis reactants will gain or lose protons in order to change the rate of reaction. Acids and bases will be studied in more depth in Big Idea #6.

A catalyst increases the rate of a chemical reaction by providing an alternative reaction pathway with a lower activation energy.

CHAPTER 6 QUESTIONS

Multiple-Choice Questions

Use the following information to answer questions 1-4

A multi-step reaction takes place with the following elementary steps:

Step I. $A + B \rightarrow C$

Step II. $C + A \rightarrow D$

Step III. $C + D \rightarrow B + E$

1. What is the overall balanced equation for this reaction?

    (A) $2A + B + 2C + D \rightarrow C + D + B + E$
    (B) $A + B \rightarrow B + E$
    (C) $A + 2C \rightarrow D + E$
    (D) $2A + C \rightarrow E$

2. What is the function of species B in this reaction?

    (A) Without it, no reaction would take place.
    (B) It is a reaction intermediate which facilitates the progress of the reaction.
    (C) It is a catalyst which changes the overall order of the reaction.
    (D) It lowers the overall activation energy of the reaction.

3. If step II is the slow step for the reaction, what is the overall rate law?

    (A) rate $= k[A]^2[C]$
    (B) rate $= k[A][C]$
    (C) rate $= k[A][B]$
    (D) rate $= k[A]/[D]$

4. Why would increasing the temperature make the reaction rate go up?

    (A) It is an endothermic reaction that needs an outside energy source to function.
    (B) The various molecules in the reactions will move faster and collide more often.
    (C) The overall activation energy of the reaction will be lowered.
    (D) A higher fraction of molecules will have the same activation energy.

5. $A + B \rightarrow C$ rate $= k[A]$

    Under which conditions would the order of the above reaction double?

    (A) doubling the initial concentration of A
    (B) doubling the initial concentration of B
    (C) reducing the concentration of A by half
    (D) changing the concentration has no effect on reaction order

6.   A + B → C + D   rate = $k[A][B]^2$

What are the potential units for the rate constant for the above reaction?

(A) s^{-1}
(B) $s^{-1}M^{-1}$
(C) $s^{-1}M^{-2}$
(D) $s^{-1}M^{-3}$

7. A multistep reaction takes place by the following mechanism:

$$A + B → C + D$$
$$A + C → D + E$$

Which of the species shown above is an intermediate in the reaction?

(A) A
(B) B
(C) C
(D) D

8.   2 NOCl → 2 NO + Cl_2

The reaction above takes place with all of the reactants and products in the gaseous phase. Which of the following is true of the relative rates of disappearance of the reactants and appearance of the products?

(A) NO appears at twice the rate that NOCl disappears.
(B) NO appears at the same rate that NOCl disappears.
(C) NO appears at half the rate that NOCl disappears.
(D) Cl_2 appears at the same rate that NOCl disappears.

9. $H_2(g) + I_2(g) → 2 HI(g)$

When the reaction given above takes place in a sealed isothermal container, the rate law is

$$Rate = k[H_2][I_2]$$

If a mole of H_2 gas is added to the reaction chamber, which of the following will be true?

(A) The rate of reaction and the rate constant will increase.
(B) The rate of reaction and the rate constant will not change.
(C) The rate of reaction will increase and the rate constant will decrease.
(D) The rate of reaction will increase and the rate constant will not change.

10.
$$A + B \rightarrow C$$

When the reaction given above takes place, the rate law is Rate = $k[A]$

If the temperature of the reaction chamber were increased, which of the following would be true?

(A) The rate of reaction and the rate constant will increase.
(B) The rate of reaction and the rate constant will not change.
(C) The rate of reaction will increase and the rate constant will decrease.
(D) The rate of reaction will increase and the rate constant will not change.

11.
$$A + B \rightarrow C$$

Based on the following experimental data, what is the rate law for the hypothetical reaction given above?

Experiment	[A] (M)	[B] (M)	Initial Rate of Formation of C (mol/L-sec)
1	0.20	0.10	3×10^{-2}
2	0.20	0.20	6×10^{-2}
3	0.40	0.20	6×10^{-2}

(A) Rate = $k[A]$
(B) Rate = $k[A]^2$
(C) Rate = $k[B]$
(D) Rate = $k[A][B]$

12.
$$A + B \rightarrow C + D$$

The rate law for the hypothetical reaction shown above is as follows:

$$Rate = k[A]$$

Which of the following changes to the system will increase the rate of the reaction?

   I. An increase in the concentration of A
  II. An increase in the concentration of B
 III. An increase in the temperature

(A) I only
(B) I and II only
(C) I and III only
(D) II and III only

13. $A + B \rightarrow C$

Based on the following experimental data, what is the rate law for the hypothetical reaction given above?

Experiment	[A] (M)	[B] (M)	Initial Rate of Formation of C (M/sec)
1	0.20	0.10	2.0×10^{-6}
2	0.20	0.20	4.0×10^{-6}
3	0.40	0.40	1.6×10^{-5}

(A) Rate = k[A]
(B) Rate = k[A]^2
(C) Rate = k[B]
(D) Rate = k[A][B]

14.

Time (Hours)	[A] M
0	0.40
1	0.20
2	0.10
3	0.05

Reactant A underwent a decomposition reaction. The concentration of A was measured periodically and recorded in the chart above. Based on the data in the chart, which of the following is the rate law for the reaction?

(A) Rate = k[A]
(B) Rate = k[A]^2
(C) Rate = $2k$[A]
(D) Rate = $\dfrac{1}{2} k$[A]

15. $$A \rightarrow B + C \qquad\qquad rate = k[A]^2$$

Which of the following graphs may have been created using data gathered from the above reaction?

(A)

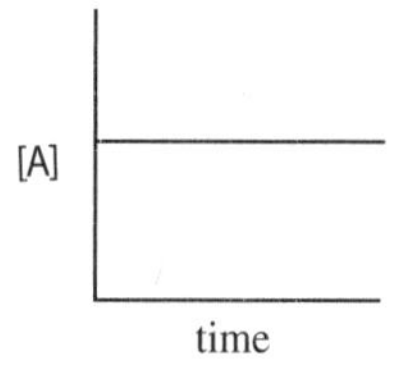

(B)

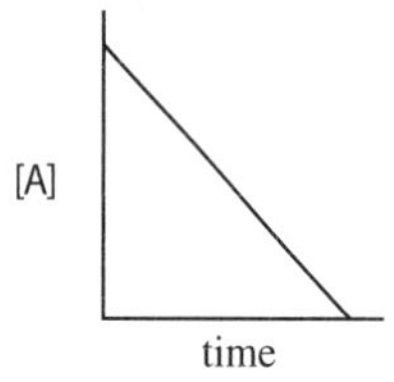

(C)

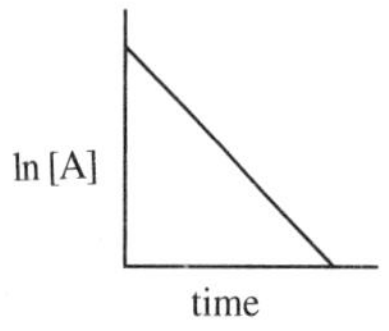

(D)

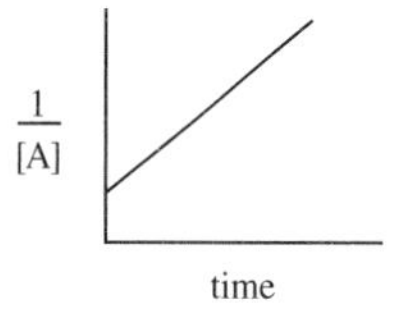

16. After 44 minutes, a sample of $^{44}_{19}K$ is found to have decayed to 25 percent of the original amount present. What is the half-life of $^{44}_{19}K$?

(A) 11 minutes
(B) 22 minutes
(C) 44 minutes
(D) 66 minutes

Constructed Response Questions

1.
$$A + 2B \rightarrow 2C$$

The following results were obtained in experiments designed to study the rate of the reaction above:

Experiment	Initial Concentration (mol/L)		Initial Rate of Disappearance of A (M/sec)
	[A]	[B]	
1	0.05	0.05	3.0×10^{-3}
2	0.05	0.10	6.0×10^{-3}
3	0.10	0.10	1.2×10^{-2}
4	0.20	0.10	2.4×10^{-2}

(a) Determine the order of the reaction with respect to each of the reactants, and write the rate law for the reaction.
(b) Calculate the value of the rate constant, k, for the reaction. Include the units.
(c) If another experiment is attempted with [A] and [B], both 0.02-molar, what would be the initial rate of disappearance of A?
(d) The following reaction mechanism was proposed for the reaction above:
$A + B \rightarrow C + D$
$D + B \rightarrow C$

  (i) Show that the mechanism is consistent with the balanced reaction.

  (ii) Show which step is the rate-determining step, and explain your choice.

2.
$$2\,NO(g) + Br_2(g) \rightarrow 2\,NOBr(g)$$

The following results were obtained in experiments designed to study the rate of the reaction above:

Experiment	Initial Concentration (mol/L)		Initial Rate of Appearance of NOBr (M/sec)
	[NO]	[Br$_2$]	
1	0.02	0.02	9.6×10^{2}
2	0.04	0.02	3.8×10^{1}
3	0.02	0.04	1.9×10^{1}

(a) Write the rate law for the reaction.
(b) Calculate the value of the rate constant, k, for the reaction. Include the units.
(c) In experiment 2, what was the concentration of NO remaining when half of the original amount of Br$_2$ was consumed?
(d) Which of the following reaction mechanisms is consistent with the rate law established in (a)? Explain your choice.

  I. $NO + NO \rightleftharpoons N_2O_2$ (fast)

  $N_2O_2 + Br_2 \rightarrow 2\,NOBr$ (slow)

  II. $Br_2 \rightarrow Br + Br$ (slow)

  $2(NO + Br \rightarrow NOBr)$ (fast)

3.
$$N_2O_5(g) \rightarrow 4\,NO_2(g) + O_2(g)$$

Dinitrogen pentoxide gas decomposes according to the equation above. The first-order reaction was allowed to proceed at 40°C and the data below were collected.

$[N_2O_5]$ (M)	Time (min)
0.400	0.0
0.289	20.0
0.209	40.0
0.151	60.0
0.109	80.0

(a) Calculate the rate constant for the reaction using the values for concentration and time given in the table. Include units with your answer.
(b) After how many minutes will $[N_2O_5]$ be equal to 0.350 M?
(c) What will be the concentration of N_2O_3 after 100 minutes have elapsed?
(d) Calculate the initial rate of the reaction. Include units with your answer.
(e) What is the half-life of the reaction?

4.
$$2\,A + B \rightarrow C + D$$

The following results were obtained in experiments designed to study the rate of the reaction above:

Experiment	Initial Concentration (moles/L)		Initial Rate of Formation of D (M/min)
	[A]	[B]	
1	0.10	0.10	1.5×10^3
2	0.20	0.20	3.0×10^3
3	0.20	0.40	6.0×10^3

(a) Write the rate law for the reaction.
(b) Calculate the value of the rate constant, k, for the reaction. Include the units.
(c) If experiment 2 goes to completion, what will be the final concentration of D? Assume that the volume is unchanged over the course of the reaction and that no D was present at the start of the experiment.
(d) Which of the following possible reaction mechanisms is consistent with the rate law found in (a)?

    I. $A + B \rightarrow C + E$ (slow)

        $A + E \rightarrow D$ (fast)

    II. $B \rightarrow C + E$ (slow)

        $A + E \rightarrow F$ (fast)

        $A + F \rightarrow D$ (fast)

(e) Calculate the half-life of reactant B.

5. $A(g) + B(g) \rightarrow C(g)$

The reaction above is second order with respect to A and zero order with respect to B. Reactants A and B are present in a closed container. Predict how each of the following changes to the reaction system will affect the rate and rate constant and explain why.

(a)   More gas A is added to the container.
(b)   More gas B is added to the container.
(c)   The temperature is increased.
(d)   An inert gas D is added to the container.
(e)   The volume of the container is decreased.

6. Use your knowledge of kinetics to answer the following questions. Justify your answers.

(a)

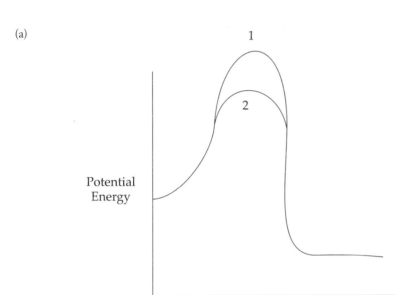

The two lines in the diagram above show different reaction pathways for the same reaction. Which of the two lines shows the reaction when a catalyst has been added?

(b)

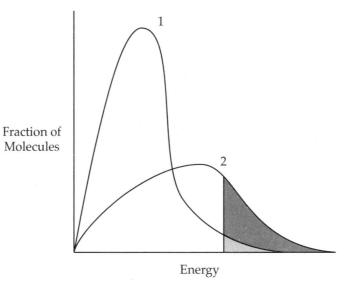

Which of the two lines in the energy distribution diagram shows the conditions at a higher temperature?

(c)

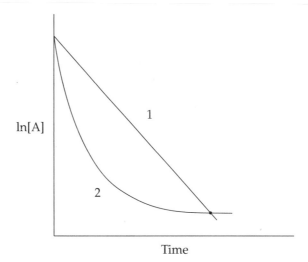

Which of the two lines in the diagram above shows the relationship of ln[A] to time for a first order reaction with the following rate law?

$$Rate = k[A]$$

(d)

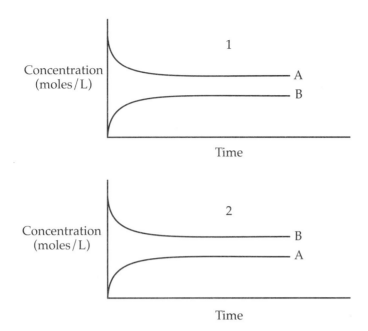

Which of the two graphs above shows the changes in concentration over time for the following reaction?

$$A \rightarrow B$$

7. Use your knowledge of kinetics to explain each of the following statements:

  (a) An increase in the temperature at which a reaction takes place causes an increase in reaction rate.
  (b) The addition of a catalyst increases the rate at which a reaction will take place.
  (c) A catalyst that has been ground into powder will be more effective than a solid block of the same catalyst.
  (d) Increasing the concentration of reactants increases the rate of a reaction.

8. The reaction between crystal violet (a complex organic molecule represented by CV^+) and sodium hydroxide is as follows:

$$CV^+ \quad + \quad OH^- \quad \rightarrow \quad CVOH$$
$$\text{(violet)} \qquad \text{(colorless)} \qquad \text{(colorless)}$$

As the crystal violet is the only colored species in the reaction, a spectrophotometer calibrated to a specific wavelength can be used to determine its concentration over time. The following data was gathered:

[CV+] (M)	Time (s)
5.5×10^{-5}	0
3.8×10^{-5}	60
2.6×10^{-5}	120
1.8×10^{-5}	180

(a) (i) What is the rate of disappearance for crystal violet from $t = 60$ s to $t = 120$ s ?
   (ii) If the solution placed in the spectrophotometer 30 s after mixing instead of immediately after mixing, how would that affect the calculated rate of disappearance for crystal violet in part (i)?
(b) Given the path length of the cuvette is 1.00 cm and the molar absorptivity of the solution is 26000 at the wavelength of the spectrophotometer, what would the absorbance reading on the spectrophotometer be at $t = 60$ s?
(c) This reaction is known to be first order with respect to crystal violet. On the provided axes, graph a function of [CV] vs. time that will provide you with a straight line graph.

The following data was also gathered over the course of three trials:

Experiment	$[CV^+]_{init}$ (M)	$[OH^-]_{init}$ (M)	Initial rate of formation of CVOH (M/s)
1	5.5×10^{-5}	0.12	3.60×10^{-7}
2	5.5×10^{-5}	0.24	7.20×10^{-7}
3	4.1×10^{-5}	0.18	?

(d) Write the rate law for this reaction.
(e) What is the rate constant, k, for this reaction? Include units in your answer.
(f) Determine the initial rate of formation of CVOH for trial 3.

CHAPTER 6 ANSWERS AND EXPLANATIONS

Multiple-Choice Questions

1. **D** The answer to (A) shows all of the reactants and products that are part of this reaction. However, species that appear on both sides need to cancel. That eliminates B, one C, and D, leaving $2A + C \rightarrow E$

2. **D** Species B is a catalyst as it is present before before and after the reaction but cancels out of the overall equation. However, catalysts do not change the order of the reaction, they simply lower the activation energy so the reaction can proceed more quickly. The reaction would still take place without the catalyst, but it would proceed very slowly.

3. **B** The rate law for the slowest elementary step is also the overall rate law for the reaction. The rate law for any elementary step can be determined by taking the concentration of each reactant in that step and raising it to the power of any coefficient. Neither species A nor C has a coefficient in step 2, meaning they are both raised to the power of 1.

4. **B** (D) is tempting, but in a Boltzmann distribution you get a higher "peak" of molecules at the same speed when the temperature is lower. At higher temperature, overall there are more molecules above the activation energy value needed for the reaction to progress, but they are more widely distributed.

5. **D** While the rate of reaction may be increased by changing the concentration of A, the rate order, which described how changing the concentration of A effects the speed of the reaction, will not change.

6. **C** To answer this we can do a unitless dimensional analysis:

    $M/s = k(M)(M)^2$
    $M/s = k(M)^3$
    $k = s^{-1}M^{-2}$

7. **C** C is created and used up in the reaction, so it will not be present in the balanced equation.

8. **B**   For every two NO molecules that form, two NOCl molecules must disappear, so NO is appearing at the same rate that NOCl is disappearing. Choices (D) and (E) are wrong because for every mole of Cl_2 that forms, two moles of NOCl are disappearing, so Cl_2 is appearing at *half* the rate that NOCl is disappearing.

9. **D**   From the rate law given in the question (Rate = $k[H_2][I_2]$), we can see that increasing the concentration of H_2 will increase the rate of reaction. The rate constant, k, is not affected by changes in the concentration of the reactants.

10. **A**   When temperature increases, the rate constant increases to reflect the fact that more reactant molecules are likely to have enough energy to react at any given time. So both the rate constant and the rate of reaction will increase.

11. **C**   From a comparison of experiments 1 and 2, when [B] is doubled while [A] is held constant, the rate doubles. That means that the reaction is first order with respect to B.

From a comparison of experiments 2 and 3, when [A] is doubled while [B] is held constant, the rate doesn't change. That means that the reaction is zero order with respect to A and that A will not appear in the rate law.

So the rate law is Rate = $k[B]$.

12. **C**   An increase in the concentration of A will increase the rate, as shown in the rate law for the reaction, so (I) is correct. Reactant B is not included in the rate law, so an increase in the concentration of B will not affect the rate; therefore, (II) is wrong. An increase in temperature causes more collisions with greater energy among reactants and always increases the rate of a reaction, so (III) is correct.

13. **D**   From a comparison of experiments 1 and 2, when [B] is doubled while [A] is held constant, the rate doubles. That means that the reaction is first order with respect to B.

From a comparison of experiments 2 and 3, when both [A] and [B] are doubled, the rate increases by a factor of 4. We would expect the rate to double based on the change in B; because the rate is in fact multiplied by 4, the doubling of A must also change the rate by a factor of 2, so the reaction is also first order with respect to A.

So the rate law is Rate = $k[A][B]$.

14. **A** The key to this question is to recognize that reactant A is disappearing with a characteristic half-life. This is a signal that the reaction is first order with respect to A. So the rate law must be Rate = k[A].

15. **D** The reaction is a second order reaction, of which a graph of the inverse of concentration always produces a straight line with a slope equal to k, the rate constant.

16. **B** Make a chart. Always start at time = 0.

Half-Lives	Time	Stuff
0	0	100%
1	X	50%
2	44 min.	25%

It takes two half-lives for the amount of $^{44}_{19}$K to decrease to 25 percent. If two half-lives takes 44 minutes, one half-life must be 22 minutes.

Constructed Response Questions

1. (a) When we compare the results of experiments 3 and 4, we see that when [A] doubles, the rate doubles, so the reaction is first order with respect to A.

    When we compare the results of experiments 1 and 2, we see that when [B] doubles, the rate doubles, so the reaction is first order with respect to B.

    Rate = k[A][B]

    (b) Use the values from experiment 3, just because they look the simplest.

    $$k = \frac{\text{Rate}}{[\text{A}][\text{B}]} = \frac{\left(1.2 \times 10^{-2} \ M/\text{sec}\right)}{\left(0.10 \ M\right)\left(0.10 \ M\right)} = 1.2 \ M^{-1}\text{sec}^{-1} = 1.2 \ \text{L/mol-sec}$$

    (c) Use the rate law.

    Rate = k[A][B]

    Rate = $(1.2 \ M^{-1}\text{sec}^{-1})(0.02 \ M)(0.02 \ M) = 4.8 \times 10^{-4} \ M/\text{sec}$

(d) (i) $A + B \rightarrow C + D$

$D + B \rightarrow C$

The two reactions add up to

$A + 2B + D \rightarrow 2C + D$

D's cancel, and we're left with the balanced equation.

$A + 2B \rightarrow 2C$

(ii) $A + B \rightarrow C + D$ (slow)

$D + B \rightarrow C$ (fast)

The first part of the mechanism is the slow, rate-determining step because its rate law is the same as the experimentally determined rate law.

2. (a) When we compare the results of experiments 1 and 2, we see that when [NO] doubles, the rate quadruples, so the reaction is second order with respect to NO.

When we compare the results of experiments 1 and 3, we see that when [Br_2] doubles, the rate doubles, so the reaction is first order with respect to Br_2.

Rate = $k[NO]^2[Br_2]$

(b) Use the values from experiment 1, just because they look the simplest.

$$k = \frac{\text{Rate}}{[NO]^2[Br_2]} = \frac{\left(9.6 \times 10^{-2} \ M/sec\right)}{\left(0.02 \ M\right)^2 \left(0.02 \ M\right)} = 1.2 \times 10^4 \ M^{-2}sec^{-1} =$$

$1.2 \times 10^4 \ L^2/mol^2\text{-sec}$

(c) In experiment 2, we started with [Br_2] = 0.02 M, so 0.01 M was consumed.

From the balanced equation, 2 moles of NO are consumed for every mole of Br_2 consumed. So 0.02 M of NO are consumed.

[NO] remaining = 0.04 M – 0.02 M = 0.02 M

(d)   Choice (I) agrees with the rate law.

$$NO + NO \leftrightarrow N_2O_2 \qquad \text{(fast)}$$

$$N_2O_2 + Br_2 \rightarrow 2\ NOBr \qquad \text{(slow)}$$

The slow step is the rate-determining step, with the following rate law:

$$\text{Rate} = k[N_2O_2][Br_2]$$

Looking at the first step in the proposed mechanism, we can replace the N_2O_2 with 2 NO. Doing so gives us:

$$2\ NO + Br_2 \rightarrow 2\ NOBr$$

The rate law for that elementary step matches the overall rate law determined previously.

3.   (a)   Use the first-order rate law and insert the first two lines from the table.

$$\ln[N_2O_5]_t - \ln[N_2O_5]_o = -kt$$

$$\ln(0.289) - \ln(0.400) = -k(20.0\ \text{min})$$

$$-.325 = -k(20.0\ \text{min})$$

$$k = 0.0163\ \text{min}^{-1}$$

(b)   Use the first-order rate law.

$$\ln[N_2O_5]_t - \ln[N_2O_5]_o = -kt$$

$$\ln(0.350) - \ln(0.400) = -(0.0163\ \text{min}^{-1})t$$

$$-0.134 = -(0.0163\ \text{min}^{-1})t$$

$$t = 8.19\ \text{min}$$

(c)   Use the first-order rate law.

$$\ln[N_2O_5] - \ln[N_2O_5]_o = -kt$$

$$\ln[N_2O_5] - \ln(0.400) = -(0.0163\ \text{min}^{-1})(100\ \text{min})$$

$$\ln[N_2O_5] + 0.916 = 1.63$$

$$\ln[N_2O_5] = -2.55$$

$$[N_2O_5] = e^{-2.55}\ M = 0.078\ M$$

(d) For a first-order reaction, Rate = $k[N_2O_5]$ = (0.0163 min^{-1})((0.400 M) = 0.0652 M/min)

(e) You can see from the numbers on the table that the half-life is slightly over 40 min. To calculate it exactly, use the formula

$$\text{Half-life} = \frac{0.693}{k} = \frac{0.693}{0.0163 \text{ min}^{-1}} = 42.5 \text{ min}$$

4. (a) When we compare the results of experiments 2 and 3, we see that when [B] doubles, the rate doubles, so the reaction is first order with respect to B.

Experiments 2 and 3 prove that the rate must double when [B] doubles. Knowing this, we can see that when the rate doubles from experiment 1 to 2, it must be because of B, and the change in A has no effect. So the reaction is zero order with respect to A.

Rate = $k[B]$

(b) Take the values from experiment 1.

$$k = \frac{\text{Rate}}{[B]} = \frac{\left(1.5 \times 10^{-3} \text{ M/min}^{-1}\right)}{\left(0.10 \text{ M}\right)} = 1.5 \times 10^{-2} \text{min}^{-1}$$

(c) In experiment 2, there are equal concentrations of A and B, but A is consumed twice as fast, so A is the limiting reagent.

0.2 M of A is consumed.

According to the balanced equation, for every 2 moles of A consumed, 1 mole of D is produced.

So if 0.2 M of A is consumed, 0.1 M of D is produced.

(d) The reaction mechanism in choice (II) is consistent with the rate law.

B → C + E   (slow)

A + E → F   (fast)

A + F → D   (fast)

The slow, rate-determining step gives the proper rate law:

Rate = $k[B]$

By the way, choice (I) gives a rate law of Rate = $k[A][B]$

(e)   The reaction is first order with respect to B, so use the half-life formula.

$$\text{Half-Life} = \frac{0.693}{k} = \frac{0.693}{1.5 \times 10^2 \, \text{min}^{-1}} = 46 \, \text{min}$$

5.   (a)   The rate of the reaction will increase because the rate depends on the concentration of A as given in the rate law: Rate = $k[A]^2$.

The rate constant is independent of the concentration of the reactants and will not change.

(b)   The rate of the reaction will not change. If the reaction is zero order with respect to B, then the rate is independent of the concentration of B.

The rate constant is independent of the concentration of the reactants and will not change.

(c)   The rate of the reaction will increase with increasing temperature because the rate constant increases with increasing temperature.

The rate constant increases with increasing temperature because at a higher temperature more gas molecules will collide with enough energy to overcome the activation energy for the reaction.

(d)   Neither the rate nor the rate constant will be affected by the addition of an inert gas.

(e)   The rate of the reaction will increase because decreasing the volume of the container will increase the concentration of A. Rate = $k[A]^2$

The rate constant is independent of the concentration of the reactants and will not change.

6.   (a)   Line 2 is the catalyzed reaction. Adding a catalyst lowers the activation energy of the reaction, making it easier for the reaction to occur.

(b)   Line 2 shows the higher temperature distribution. At a higher temperature, more of the molecules will be at higher energies, causing the distribution to flatten out and shift to the right.

(c)   Line 1 is correct. ln[reactant] for a first order reaction changes in a linear fashion over time, as shown in the following equation.

$$\ln[A]_t = -kt + \ln[A]_o$$

$$y = mx + b$$

Notice the similarity to the slope-intercept form for a linear equation.

(d)   Graph 1 is correct, showing a decrease in the concentration of A as it is consumed in the reaction, and a corresponding increase in the concentration of B as it is produced.

7. (a)   An increase in temperature means an increase in the energy of the molecules present. If the molecules have more energy, then more of them will collide more often with enough energy to overcome the activation energy required for a reaction to take place, causing the reaction to proceed more quickly.

    (b)   A catalyst offers a reaction an alternative pathway with a lower activation energy. If the activation energy is lowered, then more molecular collisions will occur with enough energy to overcome the activation energy, causing the reaction to proceed more quickly.

    (c)   The effectiveness of a solid catalyst depends on the surface area of the catalyst that is exposed to the reactants. Grinding a solid into powder greatly increases its surface area.

    (d)   Increasing the concentration of reactants crowds the reactants more closely together, making it more likely that they will collide with one another. The more collisions that occur, the more likely that collisions that will result in reactions will occur.

8. (a) $\dfrac{\left(3.8 \times 10^{-5}\ \text{M}\right) - \left(2.6 \times 10^{-5}\ \text{M}\right)}{60\ \text{s} - 120\ \text{s}} = 2.0 \times 10^{-7}\ \text{M/s}$

    (b)   A = abc
            A = (26,000 cm/M)(1.00 cm)(3.8 × 10^{-5} M)
            A = 0.98

(c) A reaction that is first order with respect to $[CV^+]$ will create a straight line in a graph of ln $[CV^+]$ vs. time.

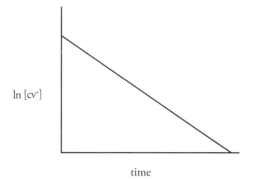

(d) Between trials 1 and 2, the value of the $[CV^+]$ remained constant while the $[OH^-]$ doubled. At the same time, the rate of reaction also doubled, meaning there is a direct relationship between $[OH^-]$ and the rate so, the reaction is first order with respect to $[OH^-]$. The rate with respect to $[CV^+]$ has already been established as first order, so:

rate = $k[CV^+][OH^-]$

(e) Either of the trials can be used for this. Taking trial 2:

7.20×10^{-7} M/s = $k(5.5 \times 10^{-5}$ M$)(0.24$ M$)$

$k = 0.055$ M^{-1}s^{-1}

(f) rate = $k[CV^+][OH^-]$
rate = $(0.055$ M^{-1}s$^{-1})(4.1 \times 10^{-5}$ M$)(0.18$ M$)$
rate = 4.1×10^{-7} s^{-1}

Chapter 7
Big Idea #5: Laws of Thermodynamics and Changes in Matter

The laws of thermodynamics describe the essential role of energy and explain and predict the direction of changes in matter.

ENERGY VS TEMPERATURE

Students often make the mistake of confusing heat and temperature. Although the two terms are interchangeably used in everyday language, they mean two different things. Temperature is a measurement of the average kinetic energy of a substance. This can be best seen with the Boltzmann Distribution, which plots the number of particles versus the kinetic energy of the particle.

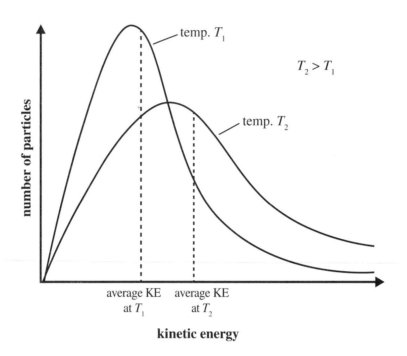

The units of temperature are °F or °R in English units, or °C or K in Metric units.

In contrast, heat is the total energy due to the molecular motions in a substance. The unit of heat is the joule in metric units. Comparing the two, it is important to remember that temperature is not energy, but a representation of the kinetic energy of the molecules.

ENERGY TRANSFER

Suppose we had two objects of different temperatures, one at 100°C and the other at 0°C, and we put the two in contact with one another. Spontaneously (i.e. it happens on its own), heat will transfer from the object with the higher temperature to the one with the lower temperature. Over time, the colder object will exhibit a temperature rise, indicating that energy has transferred from the warmer object. This is an example of heat transfer.

$$T = 100°C$$

$$T = 0°C$$

Now, suppose we have a container that is filled with liquid. The container is built in such a way that a paddle wheel mounted with a rotating shaft can mix the liquid inside. If someone outside of the container is turning the shaft, work, which is a form of energy, is being performed on the liquid. Thus, energy from outside the container is being transferred to the liquid inside the container (this can actually be verified because the temperature of the liquid will rise as the paddle wheel continues to turn). This is an example of energy transfer through work.

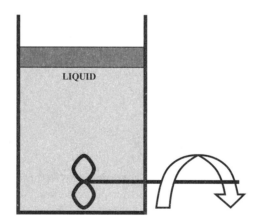

FIRST AND SECOND LAWS OF THERMODYNAMICS

The **first law of thermodynamics** says that the energy of the universe is constant. Energy can be neither created nor destroyed, so while energy can be *converted* in a chemical process, the total energy remains constant.

The **second law of thermodynamics** says that if a process is spontaneous in one direction, then it can't be spontaneous in the reverse direction, and that the entropy of the universe always increases during spontaneous reactions.

STATE FUNCTIONS

Enthalpy change (ΔH), entropy change (ΔS), and free-energy change (ΔG) are **state functions.** That means they all depend only on the change between the initial and final states of a system, not on the process by which the change occurs. For a chemical reaction, this means that the thermodynamic state functions are independent of reaction pathway; for instance, the addition of a catalyst to a reaction will have no effect on the overall energy or entropy change of the reaction.

Standard State Conditions

- All gases are at 1 atmosphere pressure.
- All liquids are pure.
- All solids are pure.
- All solutions are at 1-molar (1 M) concentration.
- The energy of formation of an element in its normal state is defined as zero.
- The temperature used for standard state values is almost invariably room temperature: 25°C (298 K). Standard state values can be calculated for other temperatures, however.

Standard State Conditions

When the values of thermodynamic quantities are given on the test, they are almost always given for standard state conditions. A thermodynamic quantity under standard state conditions is indicated by the little superscript circle, so the following is true under standard state conditions:

$$\Delta H = \Delta H^\circ$$
$$\Delta S = \Delta S^\circ$$
$$\Delta G = \Delta G^\circ$$

HEAT OF FORMATION, ΔH°_f

Heat of formation is the change in energy that takes place when one mole of a compound is formed from its component pure elements under standard state conditions. Heat of formation is almost always calculated at a temperature of 25°C (298 K).

Remember: ΔH°_f for a pure element is defined as zero.

- If ΔH°_f for a compound is negative, energy is released when the compound is formed from pure elements, and the product is *more* stable than its constituent elements. That is, the process is exothermic.

- If ΔH°_f for a compound is positive, energy is absorbed when the compound is formed from pure elements, and the product is *less* stable than its constituent elements. That is, the process is endothermic.

If the ΔH°_f values of the products and reactants are known, ΔH for a reaction can be calculated.

$$\Delta H^\circ = \Sigma\, \Delta H^\circ_f \text{ products} - \Sigma\, \Delta H^\circ_f \text{ reactants}$$

Let's find ΔH° for the reaction below.

$$2\, CH_3OH(g) + 3\, O_2(g) \rightarrow 2\, CO_2(g) + 4\, H_2O(g)$$

Compound	ΔH° (kJ/mol)
$CH_3OH(g)$	−201
$O_2(g)$	0
$CO_2(g)$	−394
$H_2O(g)$	−242

$\Delta H^\circ = \Sigma\, \Delta H^\circ_f \text{ products} - \Sigma\, \Delta H^\circ_f \text{ reactants}$
$\Delta H^\circ = [(2)(\Delta H^\circ_f CO_2) + (4)(\Delta H^\circ_f H_2O)] - [(2)(\Delta H^\circ_f CH_3OH) + (3)(\Delta H^\circ_f O_2)]$
$\Delta H^\circ = [(2)(-394 \text{ kJ}) + (4)(-242 \text{ kJ})] - [(2)(-201 \text{ kJ}) + (3)(0 \text{ kJ})]$
$\Delta H^\circ = (-1{,}756 \text{ kJ}) - (-402 \text{ kJ})$
$\Delta H^\circ = -1{,}354 \text{ kJ}$

BOND ENERGY

Bond energy is the energy required to break a bond. Because the breaking of a bond is an endothermic process, bond energy is always a positive number. When a bond is formed, energy equal to the bond energy is released.

$$\Delta H^\circ = \Sigma\, \text{ Bond energies of bonds broken} - \Sigma\, \text{ Bond energies of bonds formed}$$

The bonds broken will be the reactant bonds, and the bonds formed will be the product bonds.

Let's find $\Delta H°$ for the reaction below.

$$2\, H_2(g) + O_2(g) \rightarrow 2\, H_2O(g)$$

Bond	Bond Energy (kJ/mol)
H–H	436
O–O	499
O–H	463

$\Delta H° = \Sigma$ Bond energies of bonds broken $- \Sigma$ Bond energies of bonds formed

$\Delta H° = [(2)(H–H) + (1)(O=O)] - [(4)(O–H)]$

$\Delta H° = [(2)(436\ kJ) + (1)(499\ kJ)] - [(4)(463\ kJ)]$

$\Delta H° = (1{,}371\ kJ) - (1{,}852\ kJ)$

$\Delta H° = -481\ kJ$

HESS'S LAW

Hess's law states that if a reaction can be described as a series of steps, then ΔH for the overall reaction is simply the sum of the ΔH values for all the steps.

When manipulating equations for use in enthalpy calculations, there are two rules:

1. If you flip the equation, flip the sign on the enthalpy value
2. If you multiply or divide an equation by an integer, also multiply/divide the enthalpy value by that same integer

For example, let's say you want to calculate the enthalpy change for the following reaction:

$$4\, NH_3\,(g) + 5\, O_2\,(g) \rightarrow 4\, NO\,(g) + 6\, H_2O\,(g)$$

Given:

$N_2\,(g) + O_2\,(g) \rightarrow 2\, NO\,(g)$	$\Delta H = 180.6\ kJ/mol$
$N_2\,(g) + 3\, H_2\,(g) \rightarrow 2\, NH_3\,(g)$	$\Delta H = -91.8\ kJ/mol$
$2H_2\,(g) + O_2\,(g) \rightarrow 2\, H_2O\,(g)$	$\Delta H = -483.7\ kJ/mol$

First, we want to make sure all of the equations have the products and reactants on the correct side. A quick glance shows us the NH_3 should end up on the left but is given to us on the right, so the second equation needs to be flipped. All other species appear to be on the side they need to be on, so we'll leave the other two reactions alone.

Next, we're going to see what coefficients we need to get to. The NO needs to have a 4 but is only 2 in equation 1, so we'll multiply that equation by two. The NH_3 also needs to have a 4 but is only 2 in equation 2, so we'll also multiply that equation by two. Finally, the H_2O needs to have a 6 but only has a 2 in equation three, so the last equation gets multiplied by three. We now have:

$$2 N_2 (g) + 2 O_2 (g) \rightarrow 4 NO (g) \qquad \Delta H = 361.2 \text{ kJ/mol}$$
$$4 NH_3 (g) \rightarrow 2 N_2 (g) + 6 H_2 (g) \qquad \Delta H = 183.6 \text{ kJ/mol}$$
$$6 H_2 (g) + 3 O_2 (g) \rightarrow 6 H_2O (g) \qquad \Delta H = -1451 \text{ kJ/mol}$$

Fortunately, the two O_2 on the products side add up to 5, and both the N_2 and H_2 cancel out completely, giving us the final equation that we wanted. So, our final value for the enthalpy will be (361.2 + 183.6 − 1451 = −906 kJ/mol)"

THERMODYNAMICS OF PHASE CHANGE

Naming the Phase Changes

Phase changes occur because of changes in temperature and/or pressure. You should be able to read the phase diagram and see how a change in pressure or temperature will affect the phase of a substance.

Some particles in a liquid or solid will have enough energy to break away from the surface and become gaseous. The pressure exerted by these molecules as they escape from the surface is called the **vapor pressure**. When the liquid or solid phase of a substance is in equilibrium with the gas phase, the pressure of the gas will be equal to the vapor pressure of the substance. As temperature increases, the vapor pressure of a liquid will increase. When the vapor pressure of a liquid increases to the point where it is equal to the surrounding atmospheric pressure, the liquid boils.

Solid to *liquid*	—	**Melting**
Liquid to *solid*	—	**Freezing**
Liquid to *gas*	—	**Vaporization**
Gas to *liquid*	—	**Condensation**
Solid to *gas*	—	**Sublimation**
Gas to *solid*	—	**Deposition**

Heat of Fusion

The **heat of fusion** is the energy that must be put into a solid to melt it. This energy is needed to overcome the forces holding the solid together. Alternatively, the heat of fusion is the heat given off by a substance when it freezes. The intermolecular forces within a solid are more stable and, therefore, have lower energy than the forces within a liquid, so energy is released in the freezing process.

Heat of Vaporization

The **heat of vaporization** is the energy that must be put into a liquid to turn it into a gas. This energy is needed to overcome the forces holding the liquid together. Alternatively, the heat of vaporization is the heat given off by a substance when it condenses. Intermolecular forces become stronger when a gas condenses; the gas becomes a liquid, which is more stable, and energy is released.

As heat is added to a substance in equilibrium, the temperature of the substance can increase or the substance can change phases, but both changes cannot occur simultaneously.

Phase Diagram for Water

You should recognize how the phase diagram for water differs from the phase diagram for most other substances.

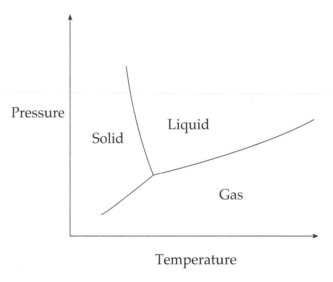

In the phase diagram for substances other than water, the solid-liquid equilibrium line slopes upward. In the phase diagram for water, the solid-liquid equilibrium line slopes *downward*. What this means is that when pressure is increased, a normal substance will change from liquid to solid, but water will change from solid to liquid.

Water has this odd property because its hydrogen bonds form a lattice structure when it freezes. This forces the molecules to remain farther apart in ice than in water, making the solid phase less dense than the liquid phase. That's why ice floats on water, and this natural phenomenon preserves most marine life.

CALORIMETRY

Heat capacity, C_p, is a measure of how much the temperature of an object is raised when it absorbs heat.

An object with a large heat capacity can absorb a lot of heat without undergoing much of a change in temperature, whereas an object with a small heat capacity shows a large increase in temperature even if only a small amount of heat is absorbed.

Specific heat is the amount of heat required to raise the temperature of one gram of a substance one degree Celsius.

Specific Heat

$$q = mc\Delta T$$

q = heat added (J or cal)
m = mass of the substance (g or kg)
c = specific heat
ΔT = temperature change (K or °C)

Calorimetry is the measurement of heat changes during chemical reactions, and is frequently accomplished via the equation above. Determining the amount of heat transfer in a reaction can lead directly to determining the enthalpy for that reaction, as shown in the example below.

$$H^+ + OH^- \rightarrow H_2O \ (l)$$

25.0 mL of 1.5 M HCl and 30.0 mL of 2.0 M NaOH are mixed together in a styrofoam cup and the reaction above occurs. The temperature of the reaction rises from 23.00°C to 31.60°C over the course of the reaction. Assuming the density of the solutions is 1.0 g/mL and the specific heat of the mixture is 4.18 J/g°C, calculate the enthalpy of the reaction.

To solve this, we first need to determine the amount of heat released during the reaction. The mass of the final solution can be determined by taking the total volume of the solution (55.0 mL) and multiplying that by the density, yielding 55.0g. The temperature change is 31.60 − 23.00 = 8.60°C. So:

$$q = mcT$$
$$q = (55.0g)(4.18J/g°C)(8.60°C)$$
$$q = 1970 \ J$$

From here, we need to determine how many moles of product are formed in this reaction. Looking at the two reactants, we can see that there is going to be less of the HCl. As everything here is in a 1:1 ratio, that means the HCl will be limiting and we can use it to determine how many moles of product will form.

$$Molarity = mol/volume$$

$$1.5 \ M = n/0.025 \ L$$

$$n = 0.038 \ mol \ HCl = 0.038 \ mol \ H_2O \ formed$$

Finally, we take the heat released and divide it by the moles of product formed to determine the overall enthalpy for this reaction.

$$1970 \text{ J}/0.038 \text{ mol} = 52,000 \text{ J/mol} = 52 \text{ kJ/mol} = \Delta H$$

It is important to emphasize that ΔH is always measured in joules per mole (or kilojoules per mole). Enthalpy is not just heat; it's the amount of heat given per mole of product that is created. Limiting reagent calculations can be done in situations where it is not easy to determine which reactant is limiting by inspection, as we did above."

ENTROPY

You should be familiar with several simple rules concerning entropies.

- Liquids have higher entropy values than solids.
- Gases have higher entropy values than liquids.
- Particles in solution have higher entropy values than solids.
- Two moles of a substance have higher entropy value than one mole.

The entropy, ΔS, of a system is a measure of the randomness or disorder of the system; the greater the disorder of a system, the greater its entropy. Because zero entropy is defined as a solid crystal at 0 K, and because 0 K has never been reached experimentally, all substances that we encounter will have some positive value for entropy. Standard entropies, $\Delta S°$, are calculated at 25°C (298 K).

The standard entropy change, $\Delta S°$, that has taken place at the completion of a reaction is the difference between the standard entropies of the products and the standard entropies of the reactants.

$$\Delta S° = \Sigma \ \Delta S°_{products} - \Sigma \ \Delta S°_{reactants}$$

GIBBS FREE ENERGY

The **Gibbs free energy**, or simply free energy, ΔG, of a process is a measure of whether or not the process will proceed without the input of outside energy. A process that occurs without outside energy input is said to be thermodynamically favored, while one that does not is said to be thermodynamically unfavored. Prior to this year's exam, the terms spontaneous and nonspontaneous were used to described thermodynamically favored and unfavored reactions, respectively. It would be good for you to be familiar with both sets of terms, although this text will use the updated terms forthwith.

For a given reaction

- if ΔG is negative, the reaction is thermodynamically favored
- if ΔG is positive, the reaction is thermodynamically unfavored
- if $\Delta G = 0$, the reaction is at equilibrium

FREE ENERGY CHANGE, ΔG

The standard free energy change, ΔG, for a reaction can be calculated from the standard free energies of formation, ΔG_f°, of its products and reactants in the same way that ΔS° was calculated.

$$\Delta G^\circ = \Sigma \; \Delta G_{f \text{ products}}^\circ - \Sigma \; \Delta G_{f \text{ reactants}}^\circ$$

ΔG, ΔH, AND ΔS

In general, nature likes to move toward two different and seemingly contradictory states—low energy and high disorder, so thermodynamically favored processes must result in decreasing enthalpy or increasing entropy or both.

There is an important equation that relates spontaneity (ΔG), enthalpy (ΔH), and entropy (ΔS) to one another.

$$\Delta G^o = \Delta H^\circ - T\Delta S^\circ$$
T = absolute temperature (K)

Note that you should make sure your units always match up here. Frequently, ΔS is given in J/mol•K and ΔH is given in kJ/mol. The convention is to convert the $T\Delta S$ term to kilojoules (kJ), as that is what Gibbs free energy is usually measured in. However, you can also convert the ΔH term to joules instead, so long as you are keeping track and labeling all units appropriately.

The chart below shows how different values of enthalpy and entropy affect spontaneity.

ΔH	ΔS	T	ΔG	
–	+	Low	–	Always spontaneous
		High	–	
+	–	Low	+	Never spontaneous
		High	+	
+	+	Low	+	Not spontaneous at low temperature
		High	–	Spontaneous at high temperature
–	–	Low	–	Spontaneous at low temperature
		High	+	Not spontaneous at high temperature

You should note that at low temperature, enthalpy is dominant, while at high temperature, entropy is dominant.

VOLTAGE AND SPONTANEITY

A redox reaction will occur spontaneously if its potential has a positive value. We also know from thermodynamics that a reaction that occurs spontaneously has a negative value for free-energy change. The relationship between reaction potential and free energy for a redox reaction is given by the equation below, which serves as a bridge between thermodynamics and electrochemistry.

$$\Delta G° = -nFE°$$

$\Delta G°$ = Standard Gibbs free energy change (kJ/mol)

n = the number of moles of electrons exchanged in the reaction (mol)

F = Faraday's constant, 96,500 coulombs/mole (that is, 1 mole of electrons has a charge of 96,500 coulombs).

$E°$ = Standard reaction potential (V)

From this equation we can see a few important things. If $E°$ is positive, $\Delta G°$ is negative and the reaction is thermodynamically favored, and if $E°$ is negative, $\Delta G°$ is positive and the reaction is thermodynamically unfavored.

Look at the reaction we saw on the previous page.

$$Zn + 2 Ag^+ \rightarrow Zn^{2+} + 2 Ag \qquad E° = +1.56 \text{ V}$$

The reaction potential is positive, so the free energy change is negative and the reaction is thermodynamically favored.

CHAPTER 7 QUESTIONS

Multiple-Choice Questions

Use the following information to answer questions 1–4.

Reaction 1: $N_2H_4 (l) + H_2 (g) \rightarrow 2\ NH_3 (g)$ $\Delta H = ?$

Reaction 2: $N_2H_4 (l) + CH_4O (l) \rightarrow CH_2O (g) + N_2 (g) + 3\ H_2 (g)$ $\Delta H = -37$ kJ/mol

Reaction 3: $N_2 (g) + 3\ H_2 (g) \rightarrow 2\ NH_3 (g)$ $\Delta H = -46$ kJ/mol

Reaction 4: $CH_4O (l) \rightarrow CH_2O (g) + H_2 (g)$ $\Delta H = -65$ kJ./mol

1. What is the enthalpy change for reaction 1?

    (A)   −148 kJ/mol

    (B)   +148 kJ/mol

    (C)   −18 kJ/mol

    (D)   −56 kJ/mol

2. Do the products or the reactants have a greater bond energy in reaction 2? Why?

    (A)   The products; this is indicated by the negative value for the enthalpy change

    (B)   The reactants; this is indicated by the negative value for the enthalpy change

    (C)   The products; this is indicated by the fact that they are less ordered than the reactants

    (D)   The reactants; this is indicated by the fact that they are less ordered than the products

3. Under what conditions would reaction 3 be thermodynamically favored?

    (A)   It is always favored.

    (B)   It is never favored.

    (C)   It is only favored at low temperatures.

    (D)   It is only favored at high temperatures.

4. If 64 g of CH_4O were to decompose via reaction 4, approximately how much energy would be released or absorbed?

    (A)   65 kJ of energy will be absorbed.

    (B)   65 kJ of energy will be released.

    (C)   130 kJ of energy will be absorbed.

    (D)   130 kJ of energy will be released.

5. $2\ Al(s) + 3\ Cl_2(g) \rightarrow 2\ AlCl_3(s)$

    The reaction above is not thermodynamically favored under standard conditions, but becomes thermodynamically favored as the temperature decreases toward absolute zero. Which of the following is true at standard conditions?

    (A) ΔS and ΔH are both negative.

    (B) ΔS and ΔH are both positive.

    (C) ΔS is negative, and ΔH is positive.

    (D) ΔS is positive, and ΔH is negative.

6.

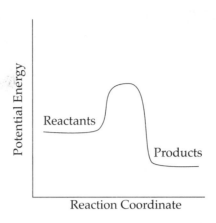

Which of the following is true of the reaction shown in the diagram above?

(A) The reaction is endothermic because the reactants are at a higher energy level than the products.
(B) The reaction is endothermic because the reactants are at a lower energy level than the products.
(C) The reaction is exothermic because the reactants are at a higher energy level than the products.
(D) The reaction is exothermic because the reactants are at a lower energy level than the products.

7. How does the addition of a catalyst affect the overall energy change of a reaction?

(A) Catalysts make reactions more endothermic as they help break the bonds of the reactants.
(B) Catalysts make reactions more exothermic as they add energy to the system.
(C) Catalysts reduce the energy change in the reaction to zero.
(D) Catalysts have no effect on the overall energy change of a reaction.

8. $C(s) + 2 H_2(g) \rightarrow CH_4(g)$ $\qquad \Delta H° = x$

$C(s) + O_2(g) \rightarrow CO_2(g)$ $\qquad \Delta H° = y$

$H_2(g) + \dfrac{1}{2} O_2(g) \rightarrow H_2O(l)$ $\qquad \Delta H° = z$

Based on the information given above, what is $\Delta H°$ for the following reaction?

$CH_4(g) + 2 O_2(g) \rightarrow CO_2(g) + 2 H_2O(l)$

(A) $x + y + z$
(B) $x + y - z$
(C) $z + y - 2x$
(D) $2z + y - x$

9. $\qquad \qquad CaCO_3 (s) \rightarrow CaO (s) + CO_2 (g)$

Is ΔS positive or negative for the above reaction, and why?

(A) Positive; the products are more disordered than the reactants
(B) Positive; the products are less ordered than the reactants
(C) Negative; the products are more disordered than the reactants
(D) Negative; the products are less ordered than the reactants

10.
$$2 H_2(g) + O_2(g) \rightarrow 2 H_2O(g)$$

Based on the information given in the table below, what is $\Delta H°$ for the above reaction?

Bond	Average bond energy (kJ/mol)
H–H	500
O=O	500
O–H	500

(A) −2,000 kJ
(B) −500 kJ
(C) +1,000 kJ
(D) +2,000 kJ

11. Which of the following is true of a reaction that is thermodynamically favored at 298 K but becomes thermodynamically unfavored at a higher temperature?

(A) $\Delta S°$ and $\Delta H°$ are both negative.
(B) $\Delta S°$ and $\Delta H°$ are both positive.
(C) $\Delta S°$ is negative, and $\Delta H°$ is positive.
(D) $\Delta S°$ is positive, and $\Delta H°$ is negative.

12. Which of the following will be true when a pure substance in liquid phase freezes spontaneously?

(A) ΔG, ΔH, and ΔS are all positive.
(B) ΔG, ΔH, and ΔS are all negative.
(C) ΔG and ΔH are negative, but ΔS is positive.
(D) ΔG and ΔS are negative, but ΔH is positive.

13.

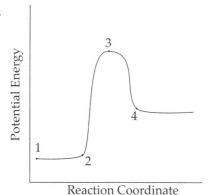

Which point on the graph shown above corresponds to activated complex or transition state?

(A) 1
(B) 2
(C) 3
(D) 4

14.

$$C(s) + O_2(g) \rightarrow CO_2(g) \qquad \Delta H° = -390 \text{ kJ/mol}$$

$$H_2(g) + \frac{1}{2} O_2(g) \rightarrow H_2O(l) \qquad \Delta H° = -290 \text{ kJ/mol}$$

$$2 C(s) + H_2(g) \rightarrow C_2H_2(g) \qquad \Delta H° = +230 \text{ kJ/mol}$$

Based on the information given above, what is $\Delta H°$ for the following reaction?

$$C_2H_2(g) + \frac{5}{2} O_2(g) \rightarrow 2 CO_2(g) + H_2O(l)$$

(A) −1,300 kJ
(B) −1,070 kJ
(C) −840 kJ
(D) −780 kJ

15. In which of the following reactions is entropy increasing?
(A) $2 SO_2(g) + O_2(g) \rightarrow 2 SO_3(g)$
(B) $CO(g) + H_2O(g) \rightarrow H_2(g) + CO_2(g)$
(C) $H_2(g) + Cl_2(g) \rightarrow 2 HCl(g)$
(D) $2 NO_2(g) \rightarrow 2 NO(g) + O_2(g)$

16. When pure sodium is placed in an atmosphere of chlorine gas, the following reaction occurs without the addition of additional energy:

$$Na (s) + Cl_2 (g) \rightarrow 2 NaCl (s)$$

Which of the following correctly identifies the values for ΔG, ΔS, and ΔH?

(A) $G > 0, S > 0, H < 0$
(B) $G < 0, S < 0, H < 0$
(C) $G < 0, S < 0, H > 0$
(D) $G > 0, S < 0, H < 0$

17.

$$H_2(g) + F_2(g) \rightarrow 2 HF(g)$$

Gaseous hydrogen and fluorine combine in the reaction above to form hydrogen fluoride with an enthalpy change of −540 kJ. What is the value of the heat of formation of $HF(g)$?

(A) −1,080 kJ/mol
(B) −270 kJ/mol
(C) 270 kJ/mol
(D) 540 kJ/mol

18.

$$H_2O (s) \rightarrow H_2O (l)$$

At room temperature, which of the following is true?

(A) G and H are positive; S is negative.
(B) G is positive; H and S are negative.
(C) G is negative; H and S are positive.
(D) G and S are negative; H is positive.

19.
$$2 S(s) + 3 O_2(g) \rightarrow 2 SO_3(g)$$
$$\Delta H = +800 \text{ kJ/mol}$$

$$2 SO_3(g) \rightarrow 2 SO_2(g) + O_2(g)$$
$$\Delta H = -200 \text{ kJ/mol}$$

Based on the information given above, what is ΔH for the following reaction?
$$S(s) + O_2(g) \rightarrow SO_2(g)$$

(A)   300 kJ
(B)   500 kJ
(C)   600 kJ
(D)   1,000 kJ

Constructed Response Questions

1.

Substance	Absolute Entropy, S (J/mol-K)	Molecular Weight
$C_6H_{12}O_6$ (s)	212.13	180
O_2 (g)	205	32
CO_2 (g)	213.6	44
H_2O (l)	69.9	18

Energy is released when glucose is oxidized in the following reaction, which is a metabolism reaction that takes place in the body.

$$C_6H_{12}O_6(s) + 6 O_2(g) \rightarrow 6 CO_2(g) + 6 H_2O(l)$$

The standard enthalpy change, $\Delta H°$, for the reaction is $-2,801$ kJ at 298 K.

(a)   Calculate the standard entropy change, $\Delta S°$, for the oxidation of glucose.
(b)   Calculate the standard free energy change, $\Delta G°$, for the reaction at 298 K.
(c)   Using the axis below, draw an energy profile for the reaction.

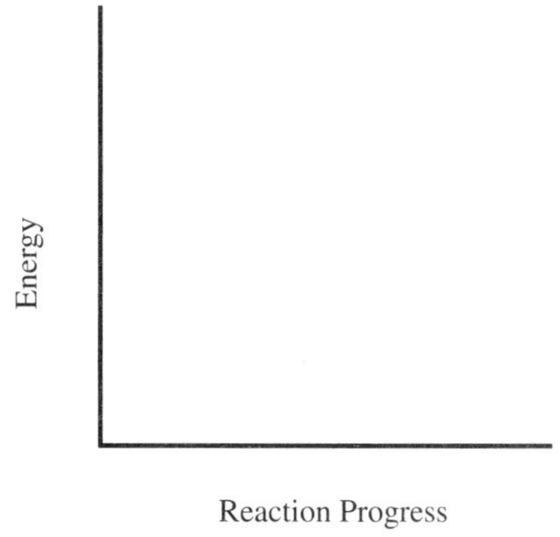

(d)   How much energy is given off by the oxidation of 1.00 gram of glucose?

2.

Bond	Average Bond Dissociation Energy (kJ/mol)
C–H	415
O=O	495
C=O	799
O–H	463

$$CH_4(g) + 2\ O_2(g) \rightarrow CO_2(g) + 2\ H_2O(g)$$

The standard free energy change, $\Delta G°$, for the reaction above is –801 kJ at 298 K.

(a) Use the table of bond dissociation energies to find $\Delta H°$ for the reaction above.
(b) How many grams of methane must react with excess oxygen in order to release 1500 kJ of heat?
(c) What is the value of $\Delta S°$ for the reaction at 298 K?
(d) Give an explanation for the size of the entropy change found in (c).

3.

$$N_2(g) + 3\ H_2(g) \rightleftharpoons 2\ NH_3(g)$$

The heat of formation, $\Delta H°_f$, of $NH_3(g)$ is – 46.2 kJ/mol. The free energy of formation, $\Delta G°_f$, of $NH_3(g)$ is –16.7 kJ/mol.

(a) What are the values of $\Delta H°$ and $\Delta G°$ for the reaction?
(b) What is the value of the entropy change, ΔS, for the reaction above at 298 K?
(c) As the temperature is increased, what is the effect on ΔG for the reaction? How does this affect the spontaneity of the reaction?
(d) At what temperature can N_2, H_2, and NH_3 gases be maintained together in equilibrium, each with a partial pressure of 1 atm?

4.

$$2\ H_2(g) + O_2(g) \rightarrow 2\ H_2O(l)$$

The reaction above proceeds spontaneously from standard conditions at 298 K.

(a) Predict the sign of the entropy change, $\Delta S°$, for the reaction. Explain.
(b) How would the value of $\Delta S°$ for the reaction change if the product of the reaction was $H_2O(g)$?
(c) What is the sign of $\Delta G°$ at 298 K? Explain.
(d) What is the sign of $\Delta H°$ at 298 K? Explain.

5.

$$CaO(s) + CO_2(g) \rightarrow CaCO_3(s)$$

The reaction above is thermodynamically favored at 298 K, and the heat of reaction, $\Delta H°$, is –178 kJ.

(a) Predict the sign of the entropy change, $\Delta S°$, for the reaction. Explain.
(b) What is the sign of $\Delta G°$ at 298 K? Explain.
(c) What change, if any, occurs to the value of $\Delta G°$ as the temperature is increased from 298 K?
(d) As the reaction takes place in a closed container, what changes will occur in the concentration of CO_2 and the temperature?

6. A student designs and experiment to determine the specific heat of aluminum. The student heats a piece of aluminum with a mass of 5.86 g to various temperatures, then drops it into a calorimeter containing 25.0 mL of water. The following data is gathered during one of the trials:

Initial Temperature of Al (°C)	Initial Temperature of H_2O (°C)	Final Temperature of Al + H_2O (°C)
109.1	23.2	26.8

(a) Given that the specific heat of water is 4.18 J/g°C and assuming its density is exactly 1.00 g/mL, calculate the heat gained by the water.

(b) Calculate the specific heat of aluminum from the experimental data given.

(c) Calculate the enthalpy of reaction for the cooling of aluminum in water in kJ/mol.

(d) If the accepted specific heat of aluminum is 0.900 J/g°C, calculate the percent error.

(e) Suggest two potential sources of error that would lead the student's experimental value to be different from the actual value. Be specific in your reasoning and make sure any identified error can be quantitatively tied to the student's results.

CHAPTER 7 ANSWERS AND EXPLANATIONS

Multiple-Choice Questions

1. **C** To get reaction 1, all that is needed is to flip reaction 4 and then add it to reactions 2 and 3. Doing so will yield reaction 1:

   $$N_2H_4 \ (l) + CH_4O \ (l) \rightarrow CH_2O \ (g) + N_2 \ (g) + 3 \ H_2 \ (g)$$
   $$N_2 \ (g) + 3 \ H_2 \ (g) \rightarrow 2 \ NH_3 \ (g)$$
   $$CH_2O \ (g) + H_2 \ (g) \rightarrow CH_4O \ (l)$$

   $$(-37 \ kJ/mol) + (-46 \ kJ/mol) + (65 \ kJ/mol) = 18 \ kJ/mol$$

2. **B** One way to calculate enthalpy is by subtracting the total bond energy of the reactants minus that of the products. A negative overall value indicates there was more bond energy in the reactants.

3. **C** Use the $\Delta G = \Delta H - T\Delta S$ equation. To do that, determine the signs of both ΔH and the $T\Delta S$ term.

   We know that ΔH is negative. ΔS is negative as well, because the reaction is going from four molecules to two molecules, meaning it is becoming more ordered. That, in turn, means the $T\Delta S$ term is positive.

   Reactions are only favored when ΔG is negative. As temperature increases, the $T\Delta S$ term becomes larger and more likely to be greater in magnitude than the ΔH term. If the temperature remains low, the $T\Delta S$ value is much more likely to be smaller in magnitude than the ΔH value, meaning ΔG is more likely to be negative and the reaction will be favored.

4. **D** The molar mass of CH_4O is 32 g/mol, so 64 g of it represent two moles. For every one mole of CH_4O that decomposes, 65 kJ of energy is released—this is indicated by the negative sign on the H value. If two moles decomposes, twice that amount—130 kJ—will be released.

5. **A** Remember that $\Delta G = \Delta H - T\Delta S$.

   If the reaction is thermodynamically favored only when the temperature is very low, then ΔG is negative only when T is very small. This can happen only when ΔH is negative (which favors spontaneity) and ΔS is negative (which favors nonspontaneity). A very small value for T will eliminate the influence of ΔS.

6. **C** In an exothermic reaction, energy is given off as the products are created because the products have less potential energy than the reactants.

7. **D** A catalyst lowers the activation energy of a reaction and makes it proceed faster. However, it has no effect on the amount of bond energy in either the reactants or products.

8. **D** The equations given above the question give the heats of formation of all the reactants and products (remember: the heat of formation of O_2, an element in its most stable form, is zero).

$\Delta H°$ for a reaction = ($\Delta H°$ for the products) – ($\Delta H°$ for the reactants).

First, the products:

> From 2 H_2O, we get $2z$
> From CO_2, we get y
> So $\Delta H°$ for the products = $2z + y$

Now the reactants:

> From CH_4, we get x. The heat of formation of O_2 is defined to be zero, so that's it for the reactants.
> $\Delta H°$ for the reaction = $(2z + y) - (x) = 2z + y - x$.

9. **A** A solid is changing to a gas, that means the products are less ordered than the reactants. Increasing disorder is represented by a positive value for ΔS.

10. **B** The bond energy is the energy that must be put into a bond to break it. First let's figure out how much energy must be put into the reactants to break their bonds.

To break 2 moles of H–H bonds, it takes (2)(500) kJ = 1,000 kJ

To break 1 mole of O=O bonds, it takes 500 kJ.

So to break up the reactants, it takes +1,500 kJ.

Energy is given off when a bond is formed; that's the negative of the bond energy. Now let's see how much energy is given off when 2 moles of H_2O are formed.

2 moles of H_2O molecules contain 4 moles of O–H bonds, so (4)(–500) kJ = –2,000 kJ are given off.

So the value of $\Delta H°$ for the reaction is

(–2,000 kJ, the energy given off) + (1,500 kJ, the energy put in) = –500 kJ.

11. A   Remember that $\Delta G = \Delta H - T\Delta S$. If the reaction is thermodynamically favored at standard temperature but becomes thermodynamically unfavored at higher temperatures, then ΔG is negative only at lower temperatures. This can happen only when ΔH is negative (which favors spontaneity) and ΔS is negative (which favors nonspontaneity). As the value of T increases, the influence of ΔS increases, eventually making the reaction thermodynamically unfavored.

12. B   The process is thermodynamically favored, so ΔG must be negative. The intermolecular forces become stronger and the substance moves to a lower energy level when it freezes, so ΔH must be negative. The substance becomes more orderly when it freezes, so ΔS must be negative.

13. C   Point 3 represents the activated complex, which is the point of highest energy. This point is the transition state between the reactants and the products.

14. A   The equations given on top give the heats of formation of all the reactants and products (remember: the heat of formation of O_2, an element in its most stable form, is zero).

$\Delta H°$ for a reaction = ($\Delta H°$ for the products) – ($\Delta H°$ for the reactants).

First, the products:

From CO_2, we get (2)(–390 kJ) = –780 kJ

From H_2O, we get –290 kJ

So $\Delta H°$ for the products = (–780 kJ) + (–290 kJ) = –1,070 kJ

Now the reactants:

From C_2H_2, we get +230 kJ. The heat of formation of O_2 is defined as zero, so that's it for the reactants.

$\Delta H°$ for the reaction = (–1,070 kJ) – (+230 kJ) = –1,300 kJ.

15. **D** Choice (D) is the only reaction where the number of moles of gas is increasing, going from 2 moles of gas on the reactant side to 3 moles of gas on the product side. In all the other choices, the number of moles of gas either decreases or remains constant.

16. **B** For the reaction to occur without the input of additional energy, it must be thermally favored, meaning $\Delta G < 0$. The product is a solid, which is more ordered than the gas which is part of the reactants, meaning $\Delta S < 0$.

    Via $\Delta G = \Delta H - T\Delta S$, if ΔS is negative, the $T\Delta S$ term is positive. Thus, for ΔG to be negative as it is in this reaction, the value for ΔH must be negative as well.

17. **B** The reaction that forms 2 moles of $HF(g)$ from its constituent elements has an enthalpy change of -540 kJ. The heat of formation is given by the reaction that forms 1 mole from these elements, so you can just divide -540 kJ by 2 to get -270 kJ.

18. **C** The reaction occurs without additional energy being added at room temperature, therefore it is favored, meaning ΔG is negative. A liquid is more disordered than a solid, meaning ΔS is positive. Finally, heat is absorbed by the ice as it melts, meaning ΔH is also positive.

19. **A** You can use Hess's law. Add the two reactions together, and cancel things that appear on both sides.

    $$2\ S(s) + 3\ O_2(g) \rightarrow 2\ SO_3(g) \qquad \Delta H = +800 \text{ kJ}$$

    $$2\ SO_3(g) \rightarrow 2\ SO_2(g) + O_2(g) \qquad \Delta H = -200 \text{ kJ}$$

    $$2\ S(s) + 3\ O_2(g) + 2\ SO_3(g) \rightarrow 2\ SO_2(g) + 2\ SO_3(g) + O_2(g)$$

    This reduces to

    $$2\ S(s) + 2\ O_2(g) \rightarrow 2\ SO_2(g) \qquad \Delta H = +600 \text{ kJ}$$

    Now we can cut everything in half to get the equation we want.

    $$S(s) + O_2(g) \rightarrow SO_2(g) \qquad \Delta H = +300 \text{ kJ}$$

Constructed Response Questions

1. (a) Use the entropy values in the table.

$$\Delta S^\circ = \Sigma S^\circ_{products} - \Sigma S^\circ_{reactants}$$

$$\Delta S^\circ = [(6)(213.6) + (6)(69.9)] - [(212.13) + (6)(205)] \text{ J/K}$$

$$\Delta S^\circ = 259 \text{ J/K}$$

(b) Use the equation below. Remember that enthalpy values are given in kJ and entropy values are given in J.

$$\Delta G^\circ = \Delta H^\circ - T \Delta S^\circ$$

$$\Delta G^\circ = (-2{,}801 \text{ kJ}) - (298)(0.259 \text{ kJ}) = -2{,}880 \text{ kJ}$$

(c)

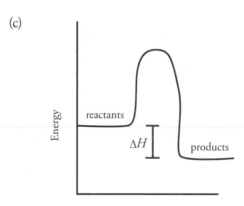

The reaction is exothermic; therefore the reactants must have more energy than the products, as indicated. The difference in energy is equal to the ΔH for this reaction.

(d) The enthalpy change of the reaction, H°, is a measure of the energy given off by 1 mole of glucose.

$$\text{Moles} = \frac{\text{grams}}{\text{MW}}$$

$$\text{Moles of glucose} = \frac{(1.00 \text{ g})}{(180 \text{ g/mol})} = 0.00556 \text{ moles}$$

$$(0.00556 \text{ mol})(2{,}801 \text{ kJ/mol}) = 15.6 \text{ kJ}$$

2. (a) Use the relationship below.

$$\Delta H^\circ = \Sigma \text{ Energies of the bonds broken } - \Sigma \text{ Energies of the bonds formed}$$

$$\Delta H^\circ = [(4)(415) + (2)(495)] - [(2)(799) + (4)(463)] \text{ kJ}$$

$$\Delta H^\circ = -800 \text{ kJ}$$

(b) $\quad 1500\,\text{kJ} \times \dfrac{1\,\text{mol}\,CH_4}{800\ \text{kJ}} = 1.9\ \text{mol}\ CH_4 \times \dfrac{16.04\ \text{g}\,CH_4}{1\,\text{mol}\,CH_4} = 30\,\text{g}\,CH_4$

(c) Use $\Delta G° = \Delta H° - T\Delta S°$

Remember that enthalpy values are given in kJ and entropy values are given in J.

$$\Delta S° = \frac{\Delta H - \Delta G}{T} = \frac{(-800\ \text{kJ}) - (-801\ \text{kJ})}{(298\ \text{K})}$$

$\Delta S° = 0.003\ \text{kJ/K} = 3\ \text{J/K}$

(d) $\Delta S°$ is very small, which means that the entropy change for the process is very small. This makes sense because the number of moles remains constant, the number of moles of gas remains constant, and the complexity of the molecules remains about the same.

3. (a) By definition, $\Delta H°_f$ and $\Delta G°_f$ for $N_2(g)$ and $H_2(g)$ are equal to zero.

$\Delta H° = \Sigma \Delta H°_f\ \text{products} - \Sigma \Delta H°_f\ \text{reactants}$

$\Delta H° = [(2\ \text{mol})(-46.2\ \text{kJ/mol})] - 0 = -92.4\ \text{kJ}$

$\Delta G° = \Sigma \Delta G°_f\ \text{products} - \Sigma \Delta G°_f\ \text{reactants}$

$\Delta G° = [(2\ \text{mol})(-16.7\ \text{kJ/mol})] - 0 = -33.4\ \text{kJ}$

(b) Use $\Delta G° = \Delta H° - T\Delta S°$

Remember that enthalpy values are given in kJ and entropy values are given in J.

$$\Delta S° = \frac{\Delta H - \Delta G}{T} = \frac{(-92.4\ \text{kJ}) - (-33.4\ \text{kJ})}{(298\ \text{K})}$$

$\Delta S° = -0.198\ \text{kJ/K} = -198\ \text{J/K}$

(c) Use $\Delta G = \Delta H° - T\Delta S°$

From (b), $\Delta S°$ is negative, so increasing the temperature increases the value of $\Delta G°$, making the reaction less thermodynamically favored.

(d) Use $\Delta G = \Delta H° - T\,\Delta S°$

At equilibrium, $\Delta G = 0$

$$T = \frac{\Delta H°}{\Delta S°} = \frac{(-92,400\ \text{J})}{(-198\ \text{J/K})} = 467\ \text{K}$$

4. (a) $\Delta S°$ is negative because the products are less random than the reactants. That's because gas is converted into liquid in the reaction.

   (b) The value of $\Delta S°$ would increase, becoming less negative because $H_2O(g)$ is more random than water but remaining negative because the entropy would still decrease from reactants to products.

   (c) $\Delta G°$ is negative because the reaction proceeds spontaneously.

   (d) $\Delta H°$ must be negative at 298 K. For a reaction to occur spontaneously from standard conditions, either $\Delta S°$ must be positive or $\Delta H°$ must be negative. This reaction is thermodynamically favored although $\Delta S°$ is negative, so $\Delta H°$ must be negative.

5. (a) $\Delta S°$ is negative because the products are less random than the reactants. That's because two moles of reactants are converted to one mole of products and gas is converted into solid in the reaction.

   (b) $\Delta G°$ is negative because the reaction proceeds spontaneously.

   (c) Use $\Delta G° = \Delta H° - T\Delta S°$

   $\Delta G°$ will become less negative because as temperature is increased, the entropy change of a reaction becomes more important in determining its spontaneity. The entropy change for this reaction is negative, which discourages spontaneity, so increasing temperature will make the reaction less thermodynamically favored, thus making $\Delta G°$ less negative.

   (d) The concentration of CO_2 will decrease as the reaction proceeds in the forward direction and the reactants are consumed. The temperature will increase as heat is given off by the exothermic reaction.

6. (a) $q = mc\Delta T$
   $q = (25.0 \text{ g})(4.18 \text{ J/g°C})(3.6 \text{ °C})$
   $q = 376 \text{ Joules}$

   (b) The heat gained by the water is the same as the heat lost by the aluminum.

   $q = mc\Delta T$
   $376 \text{ J} = (5.86\text{g})(c)(82.3°C)$
   $c = 0.780 \text{ J/g°C}$

(c) $5.86 \text{ g Al} \times \dfrac{1 \text{ mol Al}}{26.98 \text{ g Al}} = 0.217 \text{ mol Al}$

376 J/0.217 mol = 1730 J/mol = 1.73 kJ/mol

(d) $\% \text{ error} = \dfrac{\left|\text{experimental} - \text{accepted}\right|}{\text{accepted}} \times 100\%$

$\dfrac{\left|0.780 - 0.900\right|}{0.900} \times 100\% = 13.3\% \text{ error}$

(e) Error 1: If some of the heat which was lost by the aluminum was not absorbed by the water, that would cause the calculation for the heat gained by the water in part (a) be artificially low. This, in turn, would reduce the value of the specific heat of aluminum as calculated in part (b).

Error 2: If there was more than 25.0 mL of water in the calorimeter, that would mean the mass in part (a) was artificially low, which would make the calculation for the heat gained by the water also artificially low. This, in turn, would reduce the value of the specific heat of aluminum calculated in part (b).

There are many potential errors here, but as long as you can quantitatively follow them to the conclusion that the experimental value would be too low any error (within reason) can be acceptable.

Chapter 8
Big Idea #6: Equilibrium, Acids and Bases, Titrations, and Solubility

Any bond or intermolecular attraction that can be formed can be broken. These two processes are in dynamic competition, sensitive to initial conditions and external perturbations.

THE EQUILIBRIUM CONSTANT, K_{eq}

Most chemical processes are reversible. That is, reactants react to form products, but those products can also react to form reactants.

A reaction is at equilibrium when the rate of the forward reaction is equal to the rate of the reverse reaction.

The relationship between the concentrations of reactants and products in a reaction at equilibrium is given by the equilibrium expression, also called the **law of mass action.**

The Equilibrium Expression

For the reaction

$$aA + bB \rightleftharpoons cC + dD$$

$$K_{eq} = \frac{[C]^c [D]^d}{[A]^a [B]^b}$$

1. $[A]$, $[B]$, $[C]$, and $[D]$ are molar concentrations or partial pressures at equilibrium.
2. Products are in the numerator, and reactants are in the denominator.
3. Coefficients in the balanced equation become exponents in the equilibrium expression.
4. Solids and pure liquids are not included in the equilibrium expression because they cannot change their concentration. Only gaseous and aqueous species are included in the expression.
5. Units are not given for K_{eq}.

Let's look at a few examples:

1. $HC_2H_3O_2(aq) \rightleftharpoons H^+(aq) + C_2H_3O_2^-(aq)$

$$K_{eq} = K_a = \frac{[H^+][C_2H_3O_2^-]}{[HC_2H_3O_2]}$$

This reaction shows the dissociation of acetic acid in water. All of the reactants and products are aqueous particles, so they are all included in the equilibrium expression. None of the reactants or products have coefficients, so there are no exponents in the equilibrium expression. This is the standard form of K_a, the acid dissociation constant.

2. $2H_2S(g) + 3O_2(g) \rightleftharpoons 2H_2O(g) + 2SO_2(g)$

$$K_{eq} = K_c = \frac{[H_2O]^2 [SO_2]^2}{[H_2S]^2 [O_2]^3}$$

$$K_{eq} = K_p = \frac{P^2_{H_2O} P^2_{SO_2}}{P^2_{H_2S} P^3_{O_2}}$$

All of the reactants and products in this reaction are gases, so K_{eq} can be expressed in terms of concentration (K_c, moles/liter or molarity) or in terms of partial pressure (K_p, atmospheres). In the next section, we'll see how these two different ways of looking at the same equilibrium situation are related. All of the reactants and products are included here, and the coefficients in the reaction become exponents in the equilibrium expression.

3. $CaF_2(s) \rightleftharpoons Ca^{2+}(aq) + 2\ F^-(aq)$
   $K_{eq} = K_{sp} = [Ca^{2+}][F^-]^2$

This reaction shows the dissociation of a slightly soluble salt. There is no denominator in this equilibrium expression because the reactant is a solid. Solids are left out of the equilibrium expression because the concentration of a solid is constant. There must be some solid present for equilibrium to exist, but you do not need to include it in your calculations. This form of K_{eq} is called the solubility product, K_{sp}.

4. $NH_3(aq) + H_2O(l) \rightleftharpoons NH_4^+(aq) + OH^-(aq)$

$$K_{eq} = K_b = \frac{[NH_4^+][OH^-]}{[NH_3]}$$

This is the acid-base reaction between ammonia and water. We can leave water out of the equilibrium expression because it is a pure liquid. This is the standard form for K_b, the base dissociation constant.

Here is a roundup of the equilibrium constants you need to be familiar with for the test.

- K_c is the constant for molar concentrations.
- K_p is the constant for partial pressures.
- K_{sp} is the solubility product, which has no denominator because the reactants are solids.
- K_a is the acid dissociation constant for weak acids.
- K_b is the base dissociation constant for weak bases.
- K_w describes the ionization of water ($K_w = 1 \times 10^{-14}$).

A large value for K_{eq} means that products are favored over reactants at equilibrium, while a small value for K_{eq} means that reactants are favored over products at equilibrium.

The equilibrium constant has a lot of aliases, but they all take the same form and tell you the same thing. The **equilibrium constant** tells you the relative amounts of products and reactants at equilibrium.

LE CHÂTELIER'S LAW

Le Châtelier's law says that whenever a stress is placed on a situation at equilibrium, the equilibrium will shift to relieve that stress.

Let's use the **Haber process,** which is used in the industrial preparation of ammonia, as an example.

$$N_2(g) + 3\ H_2(g) \rightleftharpoons 2\ NH_3(g) \qquad\qquad \Delta H° = -92.6 \text{ kJ}$$

Concentration

When the concentration of a reactant or product is increased, the reaction will proceed in the direction that will use up the added substance.

If N_2 or H_2 is added, the reaction proceeds in the forward direction. If NH_3 is added, the reaction proceeds in the reverse direction.

When the concentration of a reactant or product is decreased, the reaction will proceed in the direction that will produce more of the substance that has been removed.

If N_2 or H_2 is removed, the reaction will proceed in the reverse direction. If NH_3 is removed, the reaction will proceed in the forward direction.

Volume

When the volume in which a reaction takes place is increased, the reaction will proceed in the direction that produces more moles of gas.

When the volume for the Haber process is increased, the reaction proceeds in the reverse direction because the reactants have more moles of gas (4) than the products (2).

When the volume in which a reaction takes place is decreased, the reaction will proceed in the direction that produces fewer moles of gas.

When the volume for the Haber process is decreased, the reaction proceeds in the forward direction because the products have fewer moles of gas (2) than the reactants (4).

If there is no gas involved in the reaction, or if the reactants and products have the same number of moles of gas, then volume changes have no effect on the equilibrium.

Temperature

There's a trick to figure out what happens when the temperature changes. First, rewrite the equation to include the heat energy on the side that it would be present on. The Haber process is exothermic, so heat is generated:

$$N_2 \, (g) + 3 \, H_2 \, (g) \leftrightarrow 2 \, NH_3 \, (g) + \text{energy}$$

Then, if the temperature goes up, the reaction will proceed in the backward direction (shifting away from the added energy). If the temperature goes down, the reaction will proceed in the forward direction (creating more energy). The reverse would be true in an endothermic reaction, as the energy would be part of the products.

Pressure

When pressure is increased, the reaction will proceed toward the side with the fewest molecules of gas.

When pressure is decreased, the reaction will proceed toward the side with the greatest number of molecules of gas.

When an inert gas, like argon, is added to the system, although the total pressure goes up, the partial pressures of the gases involved in the reaction will remain the same. Therefore, there is no effect on the equilibrium position, because the reaction quotient, Q, is determined by the individual partial pressures of the gases involved in the reaction.

THE REACTION QUOTIENT, Q

The reaction quotient is determined in exactly the same way as the equilibrium constant, but initial conditions are used in place of equilibrium conditions. The reaction quotient can be used to predict the direction in which a reaction will proceed from a given set of initial conditions.

The Reaction Quotient

For the reaction

$$aA + bB \rightleftharpoons cC + dD$$

$$Q = \frac{[C]^c [D]^d}{[A]^a [B]^b}$$

[A], [B], [C], and [D] are initial molar concentrations or partial pressures.

- If Q is less than the calculated K for the reaction, the reaction proceeds forward, generating products.
- If Q is greater than K, the reaction proceeds backward, generating reactants.
- If $Q = K$, the reaction is already at equilibrium.

The following reaction takes place in a sealed flask:

$$2\,CH_4\,(g) \leftrightarrow C_2H_2\,(g) + 3\,H_2\,(g)$$

Determine the equilibrium constant if the following concentrations are found at equilibrium.

$[CH_4] = 0.0032\ M$ $\qquad$ $[C_2H_2] = 0.025\ M$ $\qquad$ $[H_2] = 0.040\ M$

To solve this, we first need to create the equilibrium constant expression:

$$K_c = \frac{[C_2H_2][H_2]^3}{[CH_4]^2}$$

Plugging the numbers in, we get:

$$K_c = \frac{(0.025)(0.040)^3}{(0.0032)^2} \qquad K_c = \frac{1.6 \times 10^{-6}}{1.0 \times 10^{-5}} \qquad K_c = 0.16$$

Upon testing this reaction at another point, the following concentrations are found:

$[CH_4] = 0.0055\ M$ $\qquad$ $[C_2H_2] = 0.026\ M$ $\qquad$ $[H_2] = 0.029\ M$

Use the reaction quotient to determine which way the reaction needs to shift to react equilibrium.

Q uses the exact same ratios are the equilibrium constant expression, so:

$$Q = \frac{[C_2H_2][H_2]^3}{[CH_4]^2} = \frac{(0.026)(0.029)^3}{(0.0055)^2} = \frac{6.3 \times 10^{-7}}{3.0 \times 10^{-5}} = 0.21$$

0.021 is less than the equilibrium constant value of 0.16, so the reaction must proceed to the right, creating more products and reducing the amount of reactant in order to come to equilibrium.

K_{eq} AND MULTISTEP PROCESSES

There is a simple relationship between the equilibrium constants for the steps of a multistep reaction and the equilibrium constant for the overall reaction.

If two reactions can be added together to create a third reaction, then the K_{eq} for the two reactions can be multiplied together to get the K_{eq} for the third reaction.

If	$A + B \rightleftharpoons C$	$K_{eq} = K_1$
and	$C \rightleftharpoons D + E$	$K_{eq} = K_2$
then	$A + B \rightleftharpoons D + E$	$K_{eq} = K_1K_2$

SOLUBILITY

Roughly speaking, a salt can be considered "soluble" if more than 1 gram of the salt can be dissolved in 100 milliliters of water. Soluble salts are usually assumed to dissociate completely in aqueous solution. Most, but not all, solids become more soluble in a liquid as the temperature is increased.

Solubility Product (K_{sp})

Salts that are "slightly soluble" and "insoluble" still dissociate in solution to some extent. The solubility product (K_{sp}) is a measure of the extent of a salt's dissociation in solution. The K_{sp} is one of the forms of the equilibrium expression. The greater the value of the solubility product for a salt, the more soluble the salt.

Solubility Product

For the reaction
$A_aB_b(s) \rightleftharpoons a\,A^{b+}(aq) + b\,B^{a-}(aq)$

The solubility expression is
$K_{sp} = [A^{b+}]^a[B^{a-}]^b$

Here are some examples:

$$CaF_2(s) \Leftrightarrow Ca^{2+}(aq) + 2\ F^-(aq) \qquad K_{sp} = [Ca^{2+}][F^-]^2$$

$$Ag_2CrO_4(s) \Leftrightarrow 2\ Ag^+(aq) + CrO_4^{2-}(aq) \qquad K_{sp} = [Ag^+]^2[CrO_4^{2-}]$$

$$CuI(s) \Leftrightarrow Cu^+(aq) + I^-(aq) \qquad K_{sp} = [Cu^+][I^-]$$

The Common Ion Effect

Let's look at the solubility expression for AgCl.

$$K_{sp} = [Ag^+][Cl^-] = 1.6 \times 10^{-10}$$

If we throw a block of solid AgCl into a beaker of water, we can tell from the K_{sp} what the concentrations of Ag^+ and Cl^- will be at equilibrium. For every unit of AgCl that dissociates, we get one Ag^+ and one Cl^-, so we can solve the equation above as follows:

$$[Ag^+][Cl^-] = 1.6 \times 10^{-10}$$

$$(x)(x) = 1.6 \times 10^{-10}$$

$$x^2 = 1.6 \times 10^{-10}$$

$$x = [Ag^+] = [Cl^-] = 1.3 \times 10^{-5}\ M$$

So there are very small amounts of Ag^+ and Cl^- in the solution.

Let's say we add 0.10 mole of NaCl to 1 liter of the AgCl solution. NaCl dissociates completely, so that's the same thing as adding 1 mole of Na^+ ions and 1 mole of Cl^- ions to the solution. The Na^+ ions will not affect the AgCl equilibrium, so we can ignore them; but the Cl^- ions must be taken into account. That's because of the **common ion effect**.

The common ion effect says that the newly added Cl^- ions will affect the AgCl equilibrium, although the newly added Cl^- ions did not come from AgCl.

Let's look at the solubility expression again. Now we have 0.10 mole of Cl^- ions in 1 liter of the solution, so $[Cl^-] = 0.10$ M.

$$[Ag^+][Cl^-] = 1.6 \times 10^{-10}$$

$$[Ag^+](0.10\ M) = 1.6 \times 10^{-10}$$

$$[Ag^+] = \frac{\left(1.6 \times 10^{-10}\right)}{(0.10)}\ M$$

$$[Ag^+] = 1.6 \times 10^{-9}\ M$$

Now the number of Ag^+ ions in the solution has decreased drastically because of the Cl^- ions introduced to the solution by NaCl. So when solutions of AgCl and NaCl, which share a common Cl^- ion, are mixed, the more soluble salt (NaCl) can cause the less soluble salt (AgCl) to precipitate. In general, when two salt solutions that share a common ion are mixed, the salt with the lower value for K_{sp} will precipitate first.

Standard Free Energy Change and the Equilibrium Constant

The amount of Gibbs free energy in any given reaction can also be calculated if you know the equilibrium constant expression for that reaction.

$$\Delta G° = -RT \ln K$$

$\Delta G°$ = the gas constant, 8.31 J/mol-K
T = absolute temperature (K)
K = the equilibrium constant

Notice that if $\Delta G°$ is negative, K must be greater than 1, and products will be favored at equilibrium. Alternatively, if $\Delta G°$ is positive, K must be less than 1, and reactants will be favored at equilibrium.

ACIDS AND BASES DEFINITIONS

Arrhenius

S. A. Arrhenius defined an acid as a substance that ionizes in water and produces hydrogen ions (H^+ ions). For instance, HCl is an acid.

$$HCl \rightarrow H^+ + Cl^-$$

He defined a base as a substance that ionizes in water and produces hydroxide ions (OH^- ions). For instance, NaOH is a base.

$$NaOH \rightarrow Na^+ + OH^-$$

Brønsted-Lowry

J. N. Brønsted and T. M. Lowry defined an acid as a substance that is capable of donating a proton, which is the same as donating an H^+ ion, and they defined a base as a substance that is capable of accepting a proton. This definition is the one that will be used most frequently on the exam.

Look at the reversible reaction below.

$$HC_2H_3O_2 + H_2O \leftrightarrow C_2H_3O_2^- + H_3O^+$$

According to Brønsted-Lowry

$HC_2H_3O_2$ and H_3O^+ are acids.

$C_2H_3O_2^-$ and H_2O are bases.

Now look at this reversible reaction.

$$NH_3 + H_2O \leftrightarrow NH_4^+ + OH^-$$

According to Brønsted-Lowry

NH_3 and OH^- are bases.

H_2O and NH_4^+ are acids.

So in each case, the species with the H^+ ion is the acid, and the same species without the H^+ ion is the base; the two species are called a **conjugate pair.** The following are the acid-base conjugate pairs in the reactions above:

$HC_2H_3O_2$ and $C_2H_3O_2^-$

NH_4^+ and NH_3

H_3O^+ and H_2O

H_2O and OH^-

Notice that water can act either as an acid or base. Any substance which has that ability is called **amphoteric.**

Lewis

G. N. Lewis focused on electrons, and his definitions are the most broad of the acid-base definitions. Lewis defined a base as an electron pair donor and an acid as an electron pair acceptor; according to Lewis's rule, all of the Brønsted-Lowry bases above are also Lewis bases, and all of the Brønsted-Lowry acids are Lewis acids.

The following reaction is exclusively a Lewis acid-base reaction:

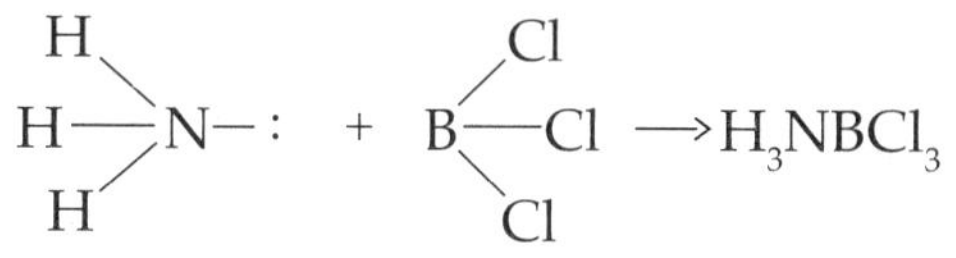

$$\underset{\underset{H}{|}}{\overset{\overset{H}{|}}{H-N-:}} \quad + \quad \underset{\underset{Cl}{|}}{\overset{\overset{Cl}{|}}{B-Cl}} \longrightarrow H_3NBCl_3$$

NH_3 is the Lewis base, donating its electron pair, and BCl_3 is the Lewis acid, accepting the electron pair.

pH

Many of the concentration measurements in acid-base problems are given to us in terms of pH and pOH.

$$p \text{ (anything)} = -\log \text{ (anything)}$$

$$pH = -\log [H^+]$$
$$pOH = -\log [OH^-]$$
$$pK_a = -\log K_a$$
$$pK_b = -\log K_b$$

In a solution

- when $[H^+] = [OH^-]$, the solution is neutral, and pH = 7
- when $[H^+]$ is greater than $[OH^-]$, the solution is acidic, and pH is less than 7
- when $[H^+]$ is less than $[OH^-]$, the solution is basic, and pH is greater than 7

It is important to remember that *increasing* pH means *decreasing* $[H^+]$, which means that there are fewer H^+ ions floating around and the solution is *less acidic*. Alternatively, *decreasing* pH means *increasing* $[H^+]$, which means that there are more H^+ ions floating around and the solution is *more acidic*.

WEAK ACIDS

When a weak acid is placed in water, a small fraction of its molecules will dissociate into hydrogen ions (H^+) and conjugate base ions (B^-). Most of the acid molecules will remain in solution as undissociated aqueous particles.

The dissociation constants, K_a and K_b, are measures of the strengths of weak acids and bases. K_a and K_b are just the equilibrium constants specific to acids and bases.

Acid Dissociation Constant

$$K_a = \frac{\left[H^+ \right]\left[A^- \right]}{[HA]}$$

$[H^+]$ = molar concentration of hydrogen ions (M)

$[A^-]$ = molar concentration of conjugate base ions (M)

$[HA]$ = molar concentration of undissociated acid molecules (M)

Base Dissociation Constant

$$K_b = \frac{\left[HA^+ \right]\left[OH^- \right]}{[B]}$$

$[HA^+]$ = protonated base ions (M)

$[OH^-]$ = molar concentration of hydroxide ions (M)

$[B]$ = unprotonated base molecules (M)

The greater the value of K_a, the greater the extent of the dissociation of the acid and the stronger the acid. The same thing goes for K_b.

If you know the K_a for an acid and the concentration of the acid, you can find the pH. For instance, let's look at 0.20-molar solution of $HC_2H_3O_2$, with $K_a = 1.8 \times 10^{-5}$.

First we set up the K_a equation, plugging in values.

$$HC_2H_3O_2 \rightarrow H^+ + C_2H_3O_2^-$$

$$K_a = \frac{\left[H^+ \right]\left[C_2H_3O_2^- \right]}{\left[HC_2H_3O_2 \right]}$$

Therefore, the ICE (Initial, Change, Equilibrium) table for the above problem is as follows:

	$[HC_2H_3O_2]$	$[H^+]$	$[C_2H_3O_2^-]$
Initial	0.20	0.0	0.0
Change	$-x$	$+x$	$+x$
Equilibrium	$0.20 - x$	x	x

Because every acid molecule that dissociates produces one H^+ and one $C_2H_3O_2^-$,

$$[H^+] = [C_2H_3O_2^-] = x$$

and because, strictly speaking, the molecules that dissociate should be subtracted from the initial concentration of $HC_2H_3O_2$, $[HC_2H_3O_2]$ should be $(0.20 \, M - x)$. In practice, however, x is almost always insignificant compared with the initial concentration of acid, so we just use the initial concentration in the calculation.

$$[HC_2H_3O_2] = 0.20 \, M$$

Now we can plug our values and variable into the K_a expression.

$$1.8 \times 10^{-5} = \frac{x^2}{0.20}$$

Solve for x.

$$x = [H^+] = 1.9 \times 10^{-3}$$

Now that we know $[H^+]$, we can calculate the pH.

$$pH = -\log [H^+] = -\log (1.9 \times 10^{-3}) = 2.7$$

This is the basic approach to solving many of the weak acid/base problems that will be on the test.

STRONG ACIDS

Strong acids dissociate completely in water, so the reaction goes to completion and they never reach equilibrium with their conjugate bases. Because there is no equilibrium, there is no equilibrium constant, so there is no dissociation constant for strong acids or bases.

Important Strong Acids
$$HCl, HBr, HI, HNO_3, HClO_4, H_2SO_4$$

Important Strong Bases
$$LiOH, NaOH, KOH, Ba(OH)_2, Sr(OH)_2$$

Because the dissociation of a strong acid goes to completion, there is no tendency for the reverse reaction to occur, which means that the conjugate base of a strong acid must be extremely weak.

Oxoacids are acids that contain oxygen. The greater the number of oxygen atoms attached to the central atom in an oxoacid, the stronger the acid. For instance, $HClO_4$ is stronger than $HClO_3$, which is stronger than $HClO_2$. That's because increasing the number of oxygen atoms that are attached to the central atom weakens the attraction that the central atom has for the H^+ ion.

It's much easier to find the pH of a strong acid solution than it is to find the pH of a weak acid solution. That's because strong acids dissociate completely, so the final concentration of H^+ ions will be the same as the initial concentration of the strong acid.

Let's look at a 0.010-molar solution of HCl.

HCl dissociates completely, so $[H^+] = 0.010\ M$
$pH = -\log [H^+] = -\log (0.010) = -\log (10^{-2}) = 2$

So you can always find the pH of a strong acid solution directly from its concentration.

K_W

Water comes to equilibrium with its ions according to the following reaction:

$$H_2O(l) \rightleftharpoons H^+(aq) + OH^-(aq) \qquad K_w = 1 \times 10^{-14} \text{ at } 25°C.$$
$$K_w = 1 \times 10^{-14} = [H^+][OH^-]$$
$$pH + pOH = 14$$

The common ion effect tells us that the hydrogen ion and hydroxide ion concentrations for any acid or base solution must be consistent with the equilibrium for the ionization of water. That is, no matter where the H^+ and OH^- ions came from, when you multiply $[H^+]$ and $[OH^-]$, you must get 1×10^{-14}. So for any aqueous solution, if you know the value of [H⁺], you can find out the value of [OH⁻] and vice versa.

The acid and base dissociation constants for conjugates must also be consistent with the equilibrium for the ionization of water.

$$K_w = 1 \times 10^{-14} = K_a K_b$$
$$pK_a + pK_b = 14$$

So if you know K_a for a weak acid, you can find K_b for its conjugate base and vice versa.

Neutralization Reactions

When an acid and a base mix, the acid will donate protons to the base in what is called a neutralization reaction. There are four different mechanisms for this, depending on the strengths of the acids and bases.

1. Strong acid + strong base

When a strong acid mixes with a strong base, both substances are dissociated completely. The only important ions in this type of reaction are the hydrogen and hydroxide ions (Even though not all bases have hydroxides, all strong bases do!).

Ex: HCl + NaOH   Net ionic: $H^+ + OH^- \leftrightarrow H_2O$ (l)

The net ionic equation for all strong acid/strong base reactions is identical—it is always the creation of water. The other ions involved in the reaction (in the example above, Cl^- and Na^+) act as spectator ions and do not take part in the reaction.

2. Strong acid + weak base

In this reaction, the strong acid (which dissociates completely), will donate a proton to the weak base. The product will be the conjugate acid of the weak base.

Ex: HCl + NH_3 Net Ionic: $H^+ + NH_3$ (aq) $\leftrightarrow NH_4^+$

3. Weak acid + strong base

In this reaction, the strong base will take as many protons as possible from the weak acid. The products are the conjugate base of the weak acid and water.

Ex: $HC_2H_3O_2$ + NaOH   Net ionic: $HC_2H_3O_2$ (aq) + $OH^- \leftrightarrow$
                                                      $C_2H_3O_2^- + H_2O$ (l)

4. Weak acid + weak base

This is a simple proton transfer reaction, in which the acid gives protons to the base.

Ex: $HC_2H_3O_2 + NH_3$ Net ionic: $HC_2H_3O_2$ (aq) + NH_3 (aq) $\leftrightarrow C_2H_3O_2^- + NH_4^+$

During neutralization reactions, the final pH of the solution is dependent on whether the excess ions at equilibrium are due to the strong acid/base or due to a weak acid/base. When a strong acid/base is in excess, the pH calculation is fairly straightforward.

Example 1: 35.0 mL of 1.5 M HCN, a weak acid ($K_a = 6.2 \times 10^{-10}$) is mixed with 25.0 mL of 2.5 M KOH. Calculate the pH of the final solution.

To do this, we can modify our ICE chart to determine what species are present at equilibrium. First, we must determine the number of moles of the reactants:

HCN = (1.5 M)(0.035 L) = 0.052 mol   KOH = (2.5 M)(0.025 L) = 0.062 mol

Then we set up our ICE chart (leaving values for water out, as it is a pure liquid)

	HCN +	OH^- $\leftrightarrow$	CN^+ +	H_2O
Initial	0.052	0.062	0	X
Change	−0.052	−0.052	+0.052	X
Equilibrium	0	0.010	0.052	X

The reaction will continue until the HCN runs out of protons to donate to the hydroxide. While there is some weak conjugate base left in solution, it is a weak base and its contribution to the pH of the solution will be irrelevant compared to the strength of the pure hydroxide ions. We now need to determine the concentration of the hydroxide ions at equilibrium. The total volume of the solution is the sum of both the acid and the base. So:

$$(35.0 \text{ mL} + 25.0 \text{ mL} = 60.0 \text{ mL})$$
$$[OH^-] = 0.010 \text{ mol}/0.060 \text{ L} = 0.17 \text{ M}$$
$$pOH = -\log (0.17) = 0.76$$
$$pH = 14 - 0.76 = 13.24$$

As you can see, it does not take much excess H^+ or OH^- to drive the pH of a solution to a fairly high or low value. Note that the K_a of the acid did not matter in this case, as the strong acid/base was in excess.

When a weak acid/base is in excess, is it easiest to use the Henderson-Hasselbalch equation:

The Henderson-Hasselbalch Equation

$$pH = pK_a + \log\frac{\left[A^-\right]}{\left[HA\right]}$$

[HA] = molar concentration of undissociated weak acid (M)

[A$^-$] = molar concentration of conjugate base (M)

$$pOH = pK_b + \log\frac{\left[HB^+\right]}{\left[B\right]}$$

[B] = molar concentration of weak base (M)

[HB$^+$] = molar concentration of conjugate acid (M)

Example 2: 25.0 mL of 1.0 M HCl is mixed with 60.0 mL of 0.50 M pyridine (C_5H_5N), a weak base ($K_b = 1.5 \times 10^{-9}$) Determine the pH of the solution.

Starting out, the number of moles of each reactant must be determined:

H$^+$ = (1.0 M)(0.025 L) = 0.025 mol H$^+$ $\qquad$ C_5H_5N = (0.50 M)(0.060 L) = 0.030 mol C_5H_5N

Then we set up the ICE chart:

	H$^+$ +	C_5H_5N ↔	HCH$_5$N$^+$
Initial	0.025	0.030	0.000
Change	−0.025	−0.025	+0.025
Equilibrium	0	0.005	0.025

The reaction continues until the H$^+$ runs out, leaving pyridine and its conjugate acid at equilibrium. Next, the new concentrations must be determined for both the weak base and its conjugate acid, as both will contribute to the final pH value. The total volume of the new solution is the sum of both the acid and base added, in this case, 25.0 + 60.0 = 85.0 mL. So:

[C_5H_5N] = 0.005 mol/0.085 L = 0.059 M $\qquad$ [HCH$_5$N$^+$]= 0.025 mol/ 0.085 L = 0.29 M

Using Henderson-Hasselbalch:

$$pOH = pK_b + \log \frac{\left[HCH_5N^+ \right]}{\left[C_5H_5N \right]}$$

$$pOH = -\log (1.5 \times 10^{-9}) + \log \frac{(0.29)}{(0.059)}$$

$$pOH = 8.8 + \log (4.9)$$

$$pOH = 8.8 + 0.69$$

$$pOH = 9.5$$

$$pH = 14 - 9.5 = 4.5$$

When a weak acid or base is in excess, the pH of the solution does not change as quickly as when a strong acid/base is in excess.

BUFFERS

A **buffer** is a solution with a very stable pH. You can add acid or base to a buffer solution without greatly affecting the pH of the solution. The pH of a buffer will also remain unchanged if the solution is diluted with water or if water is lost through evaporation.

A buffer is created by placing a large amount of a weak acid or base into a solution along with its conjugate, in the form of salt. A weak acid and its conjugate base can remain in solution together without neutralizing each other. This is called the **common ion effect**.

When both the acid and the conjugate base are together in the solution, any hydrogen ions that are added will be neutralized by the base, while any hydroxide ions that are added will be neutralized by the acid, without this having much of an effect on the solution's pH.

Let's say we have a buffer solution with concentrations of 0.20 M $HC_2H_3O_2$ and 0.50 M $C_2H_3O_2^-$. The acid dissociation constant for $HC_2H_3O_2$ is 1.8×10^{-5}. Let's find the pH of the solution.

We can just plug the values we have into the Henderson-Hasselbalch equation for acids.

$$pH = pK_a + \log \frac{\left[C_2H_3O_2^- \right]}{\left[HC_2H_3O \right]}$$

$$pH = -\log (1.8 \times 10^{-5}) + \log \frac{(0.50M)}{(0.20)}$$

$$pH = -\log (1.8 \times 10^{-5}) + \log (2.5)$$

$$pH = (4.7) + (0.40) = 5.1$$

Now let's see what happens when $[HC_2H_3O_2]$ and $[C_2H_3O_2^-]$ are both equal to 0.20 M.

$$pH = pK_a + \log \frac{\left[C_2H_3O_2^-\right]}{\left[HC_2H_3O\right]}$$

$$pH = -\log(1.8 \times 10^{-5}) + \log \frac{(0.20M)}{(0.20M)}$$

$$pH = -\log(1.8 \times 10^{-5}) + \log(1)$$

$$pH = (4.7) + (0) = 4.7$$

Notice that when the concentrations of acid and conjugate base in a solution are the same, $pH = pK_a$ (and $pOH = pK_b$). When you choose an acid for a buffer solution, it is best to pick an acid with a pK_a that is close to the desired pH. That way you can have almost equal amounts of acid and conjugate base in the solution, which will make the buffer as flexible as possible in neutralizing both added H^+ and OH^-.

POLYPROTIC ACIDS AND AMPHOTERIC SUBSTANCES

Some acids, such as H_2SO_4 and H_3PO_4, can give up more than one hydrogen ion in solution. These are called **polyprotic** acids.

Polyprotic acids are always more willing to give up their first protons than later protons. For example, H_3PO_4 gives up an H^+ ion (proton) more easily than does $H_2PO_4^-$, so H_3PO_4 is a stronger acid. In the same way, $H_2PO_4^-$ is a stronger acid than HPO_4^{2-}.

Substances that can act as either acids or bases are called **amphoteric** substances.

For instance

- $H_2PO_4^-$ can act as an acid, giving up a proton to become HPO_4^{2-}, or it can act as a base, accepting a proton to become H_3PO_4
- HSO_4^- can act as an acid, giving up a proton to become SO_4^{2-}, or it can act as a base, accepting a proton to become H_2SO_4
- H_2O can act as an acid, giving up a proton to become OH^-, or it can act as a base, accepting a proton to become H_3O^+

TITRATION

Neutralization reactions are generally performed by titration, where a base of known concentration is slowly added to an acid (or vice versa). The progress of a neutralization reaction can be shown in a titration curve. The diagram below shows the titration of a strong acid by a strong base.

> For this titration, the pH at the equivalence point is exactly 7 because the titration of a strong acid by a strong base produces a neutral salt solution.

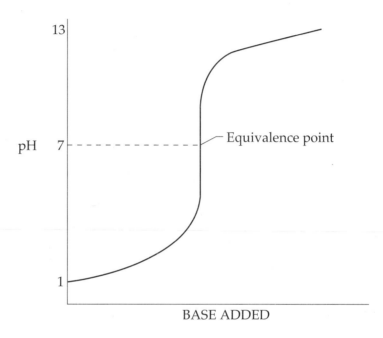

In the diagram above, the pH increases slowly but steadily from the beginning of the titration until just before the equivalence point. The **equivalence point** is the point in the titration when exactly enough base has been added to neutralize all the acid that was initially present. Just before the equivalence point, the pH increases sharply as the last of the acid is neutralized. The equivalence point of a titration can be recognized through the use of an indicator. An indicator is a substance that changes color over a specific pH range. When choosing an indicator, it's important to make sure the equivalence point falls within the pH range for the color change.

The following diagram shows the titration of a weak acid by a strong base:

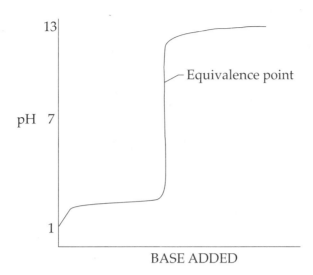

In this diagram, the pH increases more quickly at first, then levels out into a buffer region. At the center of the buffer region is the **half-equivalence point.** At this point, enough base has been added to convert exactly half of the acid into conjugate base; here the concentration of acid is equal to the concentration of conjugate base (pH = pK_a). The curve remains fairly flat until just before the equivalence point, when the pH increases sharply. For this titration, the pH at the equivalence point is greater than 7 because the titration of a weak acid by a strong base produces a basic salt solution.

The following diagram shows the titration curve of a polyprotic acid:

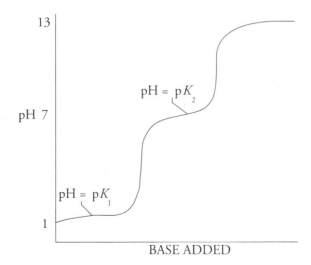

For a polyprotic acid, the titration curve will have as many bumps as there are hydrogen ions to give up. The curve above has two bumps, so it represents the titration of a diprotic acid.

CHAPTER 8 QUESTIONS

Multiple Choice

Use the following information to answer questions 1-5.

A student titrates 20.0 mL of 1.0 M NaOH with 2.0 M formic acid, HCO_2H ($K_a = 1.8 \times 10^{-4}$).

1. How much formic acid is necessary to reach the equivalence point?
   - (A) 10.0 mL
   - (B) 20.0 mL
   - (C) 30.0 mL
   - (D) 40.0 mL

2. At the equivalence point, is the solution acidic, basic, or neutral? Why?
   - (A) Acidic; the strong acid dissociates more than the weak base
   - (B) Basic; the only ion present at equilibrium is the conjugate base
   - (C) Basic; the higher concentration of the base is the determining factor
   - (D) Neutral; equal moles of both acid and base are present

3. If the formic acid were replaced with a strong acid such as HCl at the same concentration (2.0 M), how would that change the volume needed to reach the equivalence point?
   - (A) The change would reduce the amount as the acid now fully dissociates.
   - (B) The change would reduce the amount because the base will be more strongly attracted to the acid.
   - (C) The change would increase the amount because the reaction will now go to completion instead of equilibrium.
   - (D) Changing the strength of the acid will not change the volume needed to reach equivalance.

4. Which of the following would create a good buffer with the formic acid?
   - (A) $NaCO_2H$
   - (B) $HC_2H_3O_2$
   - (C) NH_3
   - (D) H_2O

5. In which pH range would a buffer made with formic acid be the most effective?
   - (A) 0-2
   - (B) 2-7
   - (C) 7-12
   - (D) 12-14

Use the following information to answer questions 6-10.

The following reaction is found to be at equilibrium at 25°C:

$$2 SO_3 (g) \leftrightarrow O_2 (g) + 2 SO_2 (g) \qquad \Delta H = -198 \text{ kJ/mol}$$

6. What is the expression for the equilibrium constant, K_c?

(A) $\dfrac{[SO_3]^2}{[O_2][SO_2]^2}$

(B) $\dfrac{2[SO_3]}{[O_2]2[SO_2]}$

(C) $\dfrac{[O_2][SO_2]^2}{[SO_3]^2}$

(D) $\dfrac{[O_2]2[SO_2]}{2[SO_3]}$

7. Which of the following would cause the reverse reaction to speed up?

(A) Adding more SO_3
(B) Raising the pressure
(C) Lowering the temperature
(D) Removing some SO_2

8. If the value for K_c at 25°C is 8.1, are the products or reactants favored, and how much?

(A) The reactants are strongly favored
(B) The reactants are weakly favored
(C) The products are strongly favored
(D) The products are weakly favored

9. Which of the following would cause a reduction in the value for the equilibrium constant?

(A) Increasing the amount of SO_3
(B) Reducing the amount of O_2
(C) Raising the temperature
(D) Lowering the temperature

10. How would raising the temperature of this reaction affects the value for ΔG. and why?

(A) ΔG would become more positive as the reaction becomes more thermodynamically favored
(B) ΔG would becomes less positive as the reaction becomes less thermodynamically favored
(C) ΔG would becomes more negative as the reaction becomes more thermodynamically favored
(D) ΔG would becomes less negative as the reaction becomes less thermodynamically favored

11. The solubility product, K_{sp}, of AgCl is 1.8×10^{-10}. Which of the following expressions is equal to the solubility of AgCl?

(A) $\left(1.8 \times 10^{-10}\right)^2$ molar

(B) $\dfrac{1.8 \times 10^{-10}}{2}$ molar

(C) 1.8×10^{-10} molar

(D) $\sqrt{1.8 \times 10^{-10}}$ molar

12. A 0.1-molar solution of which of the following acids will be the best conductor of electricity?

(A) H_2CO_3
(B) H_2S
(C) HF
(D) HNO_3

13. Which of the following expressions is equal to the K_{sp} of Ag_2CO_3?

(A) $K_{sp} = [Ag^+][CO_3^{2-}]$
(B) $K_{sp} = [Ag^+][CO_3^{2-}]^2$
(C) $K_{sp} = [Ag^+]^2[CO_3^{2-}]$
(D) $K_{sp} = [Ag^+]^2[CO_3^{2-}]^2$

14. If the solubility of BaF_2 is equal to x, which of the following expressions is equal to the solubility product, K_{sp}, for BaF_2?

(A) x^2
(B) $2x^2$
(C) $2x^3$
(D) $4x^3$

15. A beaker contains 50.0 ml of a 0.20 M Na_2SO_4 solution. If 50.0 ml of a 0.10 M solution of $Ba(NO_3)_2$ is added to the beaker, what will be the final concentration of sulfate ions in the solution?

(A) 0.20 M
(B) 0.10 M
(C) 0.050 M
(D) 0.025 M

16. The solubility of strontium fluoride in water is 1×10^{-3} M at room temperature. What is the value of the solubility product for SrF_2?

(A) 2×10^{-3}
(B) 4×10^{-6}
(C) 2×10^{-6}
(D) 4×10^{-9}

17. The bottler of a carbonated beverage dissolves carbon dioxide in water by placing carbon dioxide in contact with water at a pressure of 1 atm at room temperature. The best way to increase the amount of dissolved CO_2 would be to

(A) increase the temperature and increase the pressure of CO_2.
(B) decrease the temperature and decrease the pressure of CO_2.
(C) decrease the temperature and increase the pressure of CO_2.
(D) increase the temperature without changing the pressure of CO_2.

18. When 300. ml of a 0.60 M NaCl solution is combined with 200. ml of a 0.40 M $MgCl_2$ solution, what will be the molar concentration of Cl^- ions in the solution?

(A) 0.20 M
(B) 0.34 M
(C) 0.68 M
(D) 0.80 M

19. A student added 1 liter of a 1.0 M KCl solution to 1 liter of a 1.0 M $Pb(NO_3)_2$ solution. A lead chloride precipitate formed, and nearly all of the lead ions disappeared from the solution. Which of the following lists the ions remaining in the solution in order of decreasing concentration?

(A) $[NO_3^-] > [K^+] > [Pb^{2+}]$
(B) $[NO_3^-] > [Pb^{2+}] > [K^+]$
(C) $[K^+] > [Pb^{2+}] > [NO_3^-]$
(D) $[K^+] > [NO_3^-] > [Pb^{2+}]$

20. The solubility of PbS in water is 3×10^{-14} molar. What is the solubility product constant, K_{sp}, for PbS?

(A) 9×10^{-7}
(B) 3×10^{-14}
(C) 3×10^{-28}
(D) 9×10^{-28}

21. For a particular salt, the solution process is endothermic. As the temperature at which the salt is dissolved increases, which of the following will occur?

(A) K_{sp} will increase, and the salt will become more soluble.
(B) K_{sp} will decrease, and the salt will become more soluble.
(C) K_{sp} will increase, and the salt will become less soluble.
(D) K_{sp} will not change, and the salt will become more soluble.

22. $2\,HI(g) + Cl_2(g) \rightleftharpoons 2\,HCl(g) + I_2(g) + energy$

A gaseous reaction occurs and comes to equilibrium as shown above. Which of the following changes to the system will serve to increase the number of moles of I_2 present at equilibrium?

(A) Increasing the volume at constant temperature
(B) Decreasing the volume at constant temperature
(C) Increasing the temperature at constant volume
(D) Decreasing the temperature at constant volume

23. A sealed isothermal container initially contained 2 moles of CO gas and 3 moles of H_2 gas. The following reversible reaction occurred:

$$CO(g) + 2\,H_2(g) \rightleftharpoons CH_3OH(g)$$

At equilibrium, there was 1 mole of CH_3OH in the container. What was the total number of moles of gas present in the container at equilibrium?

(A) 1
(B) 2
(C) 3
(D) 4

24. A sample of solid potassium nitrate is placed in water. The solid potassium nitrate comes to equilibrium with its dissolved ions by the endothermic process shown below.

$$KNO_3(s) + energy \rightleftharpoons K^+(aq) + NO_3^-(aq)$$

Which of the following changes to the system would increase the concentration of K^+ ions at equilibrium?

(A) The volume of the solution is increased.
(B) The volume of the solution is decreased.
(C) Additional solid KNO_3 is added to the solution.
(D) The temperature of the solution is increased.

25. A 1 M solution of $SbCl_5$ in organic solvent shows no noticeable reactivity at room temperature. The sample is heated to 200°C and the following equilibrium reaction is found to occur:

$$SbCl_5 \rightleftharpoons SbCl_3 + Cl_2$$

The value of K_{eq} at 200°C was measured to be 10^{-6}. The reaction was then heated to 350°C, and the equilibrium concentration of Cl_2 was found to be 0.1 M, which of the following best approximated the value of K_{eq} at 350°C?

(A) 1.5
(B) 1.0
(C) 0.1
(D) 0.01

26. $2 NOBr(g) \rightleftharpoons 2 NO(g) + Br_2(g)$

The reaction above came to equilibrium at a temperature of 100°C. At equilibrium the partial pressure due to NOBr was 4 atmospheres, the partial pressure due to NO was 4 atmospheres, and the partial pressure due to Br_2 was 2 atmospheres. What is the equilibrium constant, K_p, for this reaction at 100°C?

(A) $\dfrac{1}{4}$

(B) $\dfrac{1}{2}$

(C) 1

(D) 2

27. $Br_2(g) + I_2(g) \leftrightarrow 2\ IBr(g)$

At 150°C, the equilibrium constant, K_c, for the reaction shown above has a value of 300. This reaction was allowed to reach equilibrium in a sealed container and the partial pressure due to $IBr(g)$ was found to be 3 atm. Which of the following could be the partial pressures due to $Br_2(g)$ and $I_2(g)$ in the container?

	$Br_2(g)$	$I_2(g)$
(A)	0.1 atm	0.3 atm
(B)	0.3 atm	1 atm
(C)	1 atm	1 atm
(D)	1 atm	3 atm

28. $H_2(g) + CO_2(g) \leftrightarrow H_2O(g) + CO(g)$

Initially, a sealed vessel contained only $H_2(g)$ with a partial pressure of 6 atm and $CO_2(g)$ with a partial pressure of 4 atm. The reaction above was allowed to come to equilibrium at a temperature of 700 K. At equilibrium, the partial pressure due to $CO(g)$ was found to be 2 atm. What is the value of the equilibrium constant K_p, for the reaction?

(A) $\dfrac{1}{6}$

(B) $\dfrac{1}{4}$

(C) $\dfrac{1}{3}$

(D) $\dfrac{1}{2}$

29. What is the volume of 0.05-molar HCl that is required to neutralize 50 ml of a 0.10-molar $Mg(OH)_2$ solution?

(A) 100 ml
(B) 200 ml
(C) 300 ml
(D) 400 ml

30. Which of the following best describes the pH of a 0.01-molar solution of HBrO? (For HBrO, $K_a = 2 \times 10^{-9}$)

(A) Less than or equal to 2
(B) Between 2 and 7
(C) 7
(D) Between 7 and 11

31. A laboratory technician wishes to create a buffered solution with a pH of 5. Which of the following acids would be the best choice for the buffer?

(A) $H_2C_2O_4$, $K_a = 5.9 \times 10^{-2}$

(B) H_3AsO_4, $K_a = 5.6 \times 10^{-3}$

(C) $H_2C_2H_3O_2$, $K_a = 1.8 \times 10^{-5}$

(D) HOCl, $K_a = 3.0 \times 10^{-8}$

32. Which of the following species is amphoteric?

   (A) H^+

   (B) CO_3^{2-}

   (C) HCO_3^-

   (D) H_2CO_3

33. How many liters of distilled water must be added to 1 liter of an aqueous solution of HCl with a pH of 1 to create a solution with a pH of 2?

   (A) 0.1 L
   (B) 0.9 L
   (C) 2 L
   (D) 9 L

34. A 1-molar solution of a very weak monoprotic acid has a pH of 5. What is the value of K_a for the acid?

   (A) $K_a = 1 \times 10^{-10}$
   (B) $K_a = 1 \times 10^{-7}$
   (C) $K_a = 1 \times 10^{-5}$
   (D) $K_a = 1 \times 10^{-2}$

35. The value of K_a for HSO_4^- is 1×10^{-2}. What is the value of K_b for SO_4^{2-} ?

   (A) $K_b = 1 \times 10^{-12}$
   (B) $K_b = 1 \times 10^{-8}$
   (C) $K_b = 1 \times 10^{-2}$
   (D) $K_b = 1 \times 10^2$

36. How much 0.1-molar NaOH solution must be added to 100 milliliters of a 0.2-molar H_2SO_3 solution to neutralize all of the hydrogen ions in H_2SO_3?

   (A) 100 ml
   (B) 200 ml
   (C) 300 ml
   (D) 400 ml

37. Which of the following species is amphoteric?

   (A) HNO_3
   (B) $HC_2H_3O_2$
   (C) HSO_4^-
   (D) H_3PO_4

38. If 0.630 grams of HNO_3 (molecular weight 63.0) are placed in 1 liter of distilled water at 25°C, what will be the pH of the solution? (Assume that the volume of the solution is unchanged by the addition of the HNO_3.)

   (A) 0.01
   (B) 0.1
   (C) 1
   (D) 2

39. Which of the following procedures will produce a buffered solution?

    I. Equal volumes of 0.5 M NaOH and 1 M HCl solutions are mixed.
    II. Equal volumes of 0.5 M NaOH and 1 M $HC_2H_3O_2$ solutions are mixed.
    III. Equal volumes of 1 M $NaC_2H_3O_2$ and 1 M $HC_2H_3O_2$ solutions are mixed.

(A) I only
(B) III only
(C) I and II only
(D) II and III only

40. The following titration curve shows the titration of a weak base with a strong acid:

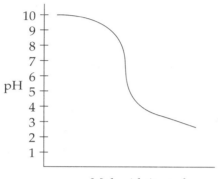

Mol acid titrated

Which of the following values most accurately approximates the pK_b of the weak base?

(A) 9.8
(B) 8.4
(C) 7
(D) 4.2

Constructed Response Questions

1. The value of the solubility product, K_{sp}, for calcium hydroxide, $Ca(OH)_2$, is 5.5×10^{-6}, at 25°C.

    (a) Write the K_{sp} expression for calcium hydroxide.
    (b) What is the mass of $Ca(OH)_2$ in 500 ml of a saturated solution at 25°C?
    (c) What is the pH of the solution in (b)?
    (d) If 1.0 mole of OH^- is added to the solution in (b), what will be the resulting Ca^{2+} concentration? Assume that the volume of the solution does not change.

2. For sodium chloride, the solution process with water is endothermic.

    (a) Describe the change in entropy when sodium chloride dissociates into aqueous particles.
    (b) Two saturated aqueous NaCl solutions, one at 20°C and one at 50°C, are compared. Which one will have higher concentration? Justify your answer.
    (c) The solubility product of $Ce_2(SO_4)_3$ decreases as temperature increases. Is the solution process for this salt endothermic or exothermic? Justify your answer.

3.
$$H_2CO_3 \rightleftharpoons H^+ + HCO_3^- \qquad\qquad K_1 = 4.3 \times 10^{-7}$$

$$HCO_3^- \rightleftharpoons H^+ + CO_3^{2-} \qquad\qquad K_2 = 5.6 \times 10^{-11}$$

The acid dissociation constants for the reactions above are given at 25°C.

(a) What is the pH of a 0.050-molar solution of H_2CO_3 at 25°C?

(b) What is the concentration of CO_3^{2-} ions in the solution in (a)?

(c) How would the addition of each of the following substances affect the pH of the solution in (a)?

    (i)   HCl

    (ii)   NaHCO$_3$

    (iii) NaOH

    (iv) NaCl

(d) What is the value of K_{eq} for the following reaction?

$$H_2CO_3 \rightleftharpoons 2\,H^+ + CO_3^{2-}$$

4. $N_2(g) + 3\,H_2(g) \rightleftharpoons 2\,NH_3(g)$ $\Delta H = -92.4$ kJ

When the reaction above took place at a temperature of 570 K, the following equilibrium concentrations were measured:

$[NH_3] = 0.20$ mol/L

$[N_2] = 0.50$ mol/L

$[H_2] = 0.20$ mol/L

(a) Write the expression for K_c and calculate its value.

(b) Calculate ΔG for this reaction

(c) Describe how the concentration of H_2 will be affected by each of the following changes to the system at equilibrium:

    (i)   The temperature is increased.

    (ii)   The volume of the reaction chamber is increased.

    (iii) N_2 gas is added to the reaction chamber.

    (iv) Helium gas is added to the reaction chamber.

5. A vessel contains 500 milliliters of a 0.100-molar H_2S solution. For H_2S, $K_1 = 1.0 \times 10^{-7}$ and $K_2 = 1.3 \times 10^{-13}$.

(a) What is the pH of the solution?

(b) How many milliliters of 0.100-molar NaOH solution must be added to the solution to create a solution with a pH of 7?

(c) What will be the pH when 800 milliliters of 0.100-molar NaOH has been added?

(d) What is the value of K_{eq} for the following reaction?

$$H_2S(aq) \leftrightarrow 2H^+(aq) + S^{2-}(aq)$$

6. A 100 milliliter sample of 0.100-molar NH_4Cl solution was added to 80 milliliters of a 0.200-molar solution of NH_3. The value of K_b for ammonia is 1.79×10^{-5}.

   (a) What is the value of pK_b for ammonia?
   (b) What is the pH of the solution described in the question?
   (c) If 0.200 grams of NaOH were added to the solution, what would be the new pH of the solution? (Assume that the volume of the solution does not change.)
   (d) If equal molar quantities of NH_3 and NH_4^+ were mixed in solution, what would be the pH of the solution?

7. A student performs an experiment to determine the concentration of a solution of hypochlorous acid, HOCl ($K_a = 3.5 \times 10^{-8}$). The student starts with 25.00 mL of the acid in a flask and titrates it against a standardized solution of sodium hydroxide with a concentration of 1.47 M. The equivalence point is reached after the addition of 34.23 mL of NaOH.

   (a) Write the net ionic equation for the reaction that occurs in the flask.
   (b) What is the concentration of the HOCl?
   (c) What would the pH of the solution in the flask be after the addition of 28.55 mL of NaOH?
   (d) The actual concentration of the HOCl is found to be 2.25 M. Quantitatively discuss whether or not each of the following errors could have caused the error in the student's results.
       (i) The student added additional NaOH past the equivalence point.
       (ii) The student rinsed the buret with distilled water but not with the NaOH solution before filling it with NaOH.
       (iii) The student measured the volume of acid incorrectly; instead of adding 25.00 mL of HOCl only 24.00 mL is present in the flask prior to titration.

CHAPTER 8 ANSWERS AND EXPLANATIONS

Multiple-Choice Questions

1. **A** The equivalence point is defined as the point at which the moles of acid is equal to the moles of base. The moles of NaOH is equal to (1.0 M)(0.0200 L) = 0.0200 mol. To calculate the volume of acid:

    2.0 M = 0.0200 mol/V

    V = 0.0100 L = 10.0 mL

2. **B** Using the ICE chart:

	HC_2OH +	OH^- ↔	O_2CH^- +	H_2O
Initial	0.020	0.020	0	X
Change	−0.020	−0.020	+ 0.020	X
Equilibrium	0	0	0.020	X

    Both the hydroxide ion and the weak acid are not present at equilibrium. The only ionic species left, the O_2CH^-, is basic, and so the solution will be as well.

3. **D** The number of moles of base is staying the same, and so is the concentration of the acid. Therfore, the same volume of that acid will be needed to get to the equivalence point

4. **A** A buffer is made up of an acid and its conjugate base. The conjugate base of HCO_2H is CO_2H^-, which is present as the anion in the $NaCO_2H$ salt.

5. **B** The ideal buffer will have a pK_a close to the pH of the buffer solution. Formic acid's K_a is 1.8×10^{-4}, meaning its pK_a will be in the 3-4 range.

6. **C** The equilibrium expression is always products over reactants, so that eliminates (A) and (B). The coefficients in the balanced equation turn into exponents in the expression, leading us to the correct answer.

7. **B** To speed up the reverse reaction, we are looking to cause a shift to the left via Le Châtelier's principle. If the pressure were to increase, that causes a shift to the side with less gas molecules, which in this case means a shift to the left. All other options cause a shift to the right.

8.   **D**   Products go in the numerator of an equilibrium expression, and as the value is bigger than one that means there will be more products than reactants at equilibrium. However, that is a fairly small value for K_c, and so while the products are favored they are not strongly favored

9.   **C**   Changing the amounts of either reactants or products present will not cause a change in the equilibrium constant (nor would changing the pressure, if that were an option). The only way to actually change the value of the constant is by changing the temperature. As this is an exothermic reaction, adding more heat would cause a shift to the left, increasing the amount of reactants, and thus the denominator in the equilibrium constant expression. This, in turn, reduces the value for K_c.

10.   **D**   First, use $\Delta G = -RT \ln K$. If the temperature of this reaction rises, it causes a shift to the left, which in turn will decrease the value of K to reduce. The value of $\ln K$ in the expression would also reduce, making the overall value for ΔG to become less negative/more positive. The answer is (D) because a shift in that direction is synonymous to a reduction in thermodynamic favoribility.

11.   **D**   The solubility of a substance is equal to its maximum concentration in solution.

For every AgCl in solution, we get one Ag^+ and one Cl^-, so the solubility of AgCl—let's call it x—will be the same as $[Ag^+]$, which is the same as $[Cl^-]$.

So for AgCl, $K_{sp} = [Ag^+][Cl^-] = 1.8 \times 10^{-10} = x^2$.

$x = \sqrt{1.8 \times 10^{-10}}$

12.   **D**   The best conductor of electricity (also called the strongest electrolyte) will be the solution that contains the most charged particles. HNO_3 is the only strong acid listed in the answer choices, so it is the only choice where the acid has dissociated completely in solution into H^+ and NO_3^- ions. So a 0.1-molar HNO_3 solution will contain the most charged particles and, therefore, be the best conductor of electricity.

13. **C** K_{sp} is just the equilibrium constant without a denominator.

When Ag_2CO_3 dissociates, we get the following reaction:

$$Ag_2CO_3(s) \rightleftharpoons 2\ Ag^+ + CO_3^{2-}$$

In the equilibrium expression, coefficients become exponents, so we get

$$K_{sp} = [Ag^+]^2[CO_3^{2-}]$$

14. **D** For BaF_2, $K_{sp} = [Ba^{2+}][F^-]^2$.

For every BaF_2 that dissolves, we get one Ba^{2+} and two F^-.

So if the solubility of BaF_2 is x, then $[Ba^{2+}] = x$, and $[F^-] = 2x$

So $K_{sp} = (x)(2x)^2 = (x)(4x^2) = 4x^3$

15. **C** The Ba^{2+} ions and the SO_4^- ions will combine and precipitate out of the solution, so let's find out how many of each we have.

Moles = (molarity)(volume)
Moles of SO_4^- = (0.20 M)(0.050 L) = 0.010 mole
Moles of Ba^{2+} = (0.10 M)(0.050 L) = 0.0050 mole

To find the number of moles of SO_4^- left in the solution, subtract the moles of Ba^{2+} from the moles of SO_4^-.

0.010 mole − 0.0050 mole = 0.0050 mole

Now use the formula for molarity to find the concentration of SO_4^- ions. Don't forget to add the volumes of the two solutions.

$$Molarity = \frac{moles}{liters} = \frac{(0.0050\ mol)}{(0.10\ L)} = 0.050\ M$$

16. **D** Use the formula for K_{sp}. For every SrF_2 in solution, there will be one strontium ion and two fluoride ions, so $[Sr^{2+}]$ will be $1 \times 10^{-3}\ M$ and $[F^-]$ will be $2 \times 10^{-3}\ M$.

$$K_{sp} = [Sr^{2+}][F^-]^2$$
$$K_{sp} = (1 \times 10^{-3}\ M)(2 \times 10^{-3}\ M)^2$$
$$K_{sp} = 4 \times 10^{-9}$$

17. **C** The lower the temperature, the more soluble a gas will be in water. The greater the pressure of the gas, the more soluble it will be.

18. **C** The Cl⁻ ions from the two salts will both be present in the solution, so we need to find the number of moles of Cl⁻ contributed by each salt.

Moles = (molarity)(volume)
Each NaCl produces 1 Cl⁻
Moles of Cl⁻ from NaCl = (0.60 M)(0.300 L) = 0.18 mole

Each MgCl₂ produces 2 Cl⁻
Moles of Cl⁻ from MgCl₂ = (2)(0.40 M)(0.200 L) = 0.16 mole

To find the number of moles of Cl⁻ in the solution, add the two together.

0.18 mole + 0.16 mole = 0.34 mole

Now use the formula for molarity to find the concentration of Cl⁻ ions. Don't forget to add the volumes of the two solutions.

$$\text{Molarity} = \frac{\text{moles}}{\text{liters}} = \frac{(0.34\,\text{mol})}{(0.500\,\text{L})} = 0.68\ \text{M}$$

19. **A** At the start, the concentrations of the ions are as follows:

$[K^+]$ = 1 M
$[Cl^-]$ = 1 M
$[Pb^{2+}]$ = 1 M
$[NO_3^-]$ = 2 M

After PbCl₂ forms, the concentrations are as follows:

$[K^+]$ = 1 M
$[Cl^-]$ = 0.5 M
$[Pb^{2+}]$ = 0 M
$[NO_3^-]$ = 2 M

So from greatest to least

$[NO_3^-] > [K^+] > [Pb^{2+}]$

20. **D** The solubility of a substance is equal to its maximum concentration in solution. For every PbS in solution, we get one Pb^{2+} and one S^{2-}, so the concentration of PbS, 3×10^{-14} M, will be the same as the concentrations of Pb^{2+} and S^{2-}.

$K_{sp} = [Pb^{2+}][S^{2-}]$
$K_{sp} = (3 \times 10^{-14}\ M)(\ 3 \times 10^{-14}\ M) = 9 \times 10^{-28}$

21. **A** From Le Châtelier's law, the equilibrium will shift to counteract any stress that is placed on it. Increasing temperature favors the endothermic direction of a reaction because the endothermic reaction absorbs the added heat. So the salt becomes more soluble, increasing the number of dissociated particles, thus increasing the value of K_{sp}.

22. **D** According to Le Châtelier's law, the equilibrium will shift to counteract any stress that is placed on it. If the temperature is decreased, the equilibrium will shift toward the side that produces energy or heat. That's the product side where I_2 is produced.

23. **C** Because the equation is balanced, the following will occur:

    If 1 mole of CH_3OH was created, then 1 mole of CO was consumed and 1 mole of CO remains; and if 1 mole of CH_3OH was created, then 2 moles of H_2 were consumed and 1 mole of H_2 remains. So at equilibrium, there are

    (1 mol CH_3OH) + (1 mol CO) + (1 mol H_2) = 3 moles of gas

24. **D** According to Le Châtelier's law, equilibrium will shift to relieve any stress placed on a system. If the temperature is increased, the equilibrium will shift to favor the endothermic reaction because it absorbs the added energy. In this case, the equilibrium will be shifted to the right, increasing the concentration of both K^+ and NO_3^- ions.

    Changing the volume of the solution, (A) and (B), will change the *number* of K^+ ions in solution, but not the *concentration* of K^+ ions. Because solids are not considered in the equilibrium expression, adding more solid KNO_3 to the solution (C) will not change the equilibrium. Decreasing the temperature (E) will favor the exothermic reaction, driving the equilibrium toward the left and decreasing the concentration of K^+ ions.

25. **D** The information about K at 200°C is useless information and can be ignored. Since all the Cl_2 found in solution must have come from $SbCl_5$, we know that at equilibrium $[Cl_2] = [SbCl_3] = 0.1$ M, and $[SbCl_5] = (1.0 - 0.1)$ M $= .99$ M. We can then say that $K = (0.1)(0.1)/.99 = .0101$ which is most closely approximated by choice D.

26. **D** $K_p = \dfrac{[NO]^2[Br_2]}{[NOBr]^2} = \dfrac{(4)^2(2)}{(4)^2} = 2$

27. **A** The equilibrium expression for the reaction is as follows:

$$\frac{P_{IBr}^{\ 2}}{P_{Br_2}P_{I_2}} = 300$$

When all of the values are plugged into the expression, (A) is the only choice that works.

$$\frac{(3)^2}{(0.1)(0.3)} = \frac{9}{0.03} = 300$$

28. **D** Use a table to see how the partial pressures change. Based on the balanced equation, we know that if 2 atm of $CO(g)$ were formed, then 2 atm of $H_2O(g)$ must also have formed. We also know that the reactants must have lost 2 atm each.

	$H_2(g)$	$CO_2(g)$	$H_2O(g)$	$CO(g)$
Before	6 atm	4 atm	0	0
Change	−2	−2	+2	+2
At Equilibrium	4 atm	2 atm	2 atm	2 atm

Now plug the numbers into the equilibrium expression.

$$K_{eq} = \frac{P_{H_2O}P_{CO}}{P_{H_2}P_{CO_2}} = \frac{(2)(2)}{(4)(2)} = \frac{1}{2}$$

29. **B** Every mole of $Mg(OH)_2$ molecules dissociates to produce 2 moles of OH^- ions, so a 0.10 M $Mg(OH)_2$ solution will be a 0.20 M OH^- solution.

The solution will be neutralized when the number of moles of H^+ ions added is equal to the number of OH^- ions originally in the solution.

Moles = (molarity)(volume)

Moles of OH^- = (0.20 M)(50 ml) = 10 millimoles = moles of H^+ added

$$Volume = \frac{moles}{molarity}$$

$$Volume\ of\ HCl = \frac{(10\ millimoles)}{(0.05\ M)} = 200\ ml$$

30. **B** You can eliminate (C), (D), and (E) by using common sense. HBrO is a weak acid, so an HBrO solution will be acidic, with a pH of less than 7.

To choose between (A) and (B) you have to remember that HBrO is a weak acid. If HBrO were a strong acid, it would dissociate completely and [H⁺] would be equal to 0.01-molar, for a pH of exactly 2. Because HBrO is a weak acid, it will not dissociate completely and [H⁺] will be less than 0.01-molar, which means that the pH will be greater than 2. So by POE, (B) is the answer.

31. **C** The best buffered solution occurs when pH = pK_a. That happens when the solution contains equal amounts of acid and conjugate base. If you want to create a buffer with a pH of 5, the best choice would be an acid with a pK_a that is as close to 5 as possible. You shouldn't have to do a calculation to see that the pK_a for choice (C) is much closer to 5 than that of any of the others.

32. **C** An amphoteric species can act either as an acid or a base, gaining or losing a proton.

   HCO_3^- can act as an acid, losing a proton to become CO_3^{2-}, or it can act as a base, gaining a proton to become H_2CO_3.

33. **D** We want to change the hydrogen ion concentration from 0.1 M (pH of 1) to 0.01 M (pH of 2).

   The HCl is completely dissociated, so the number of moles of H⁺ will remain constant as we dilute the solution.

   Moles = (molarity)(volume) = Constant

   $(M_1)(V_1) = (M_2)(V_2)$

   $(0.1\ M)(1\ L) = (0.01\ M)(V_2)$

   So, V_2 = 10 L, which means that 9 L must be added.

34. **A** A pH of 5 means that [H⁺] = 1 × 10⁻⁵

   $$K_a = \frac{[H^+][A^-]}{[HA]}$$

   For every HA that dissociates, we get one H⁺ and one A⁻, so [H⁺] = [A⁻] = 1 × 10⁻⁵.

   The acid is weak, so we can assume that very little HA dissociates and that the concentration of HA remains 1-molar.

   $$K_a = \frac{[H^+][A^-]}{[HA]} = \frac{(1 \times 10^{-5})(1 \times 10^{-5})}{(1)} = 1 \times 10^{-10}.$$

35. **A** For conjugates, $(K_a)(K_b) = K_w = 1 \times 10^{-14}$

$$K_b = \frac{K_w}{K_a} = \frac{\left(1 \times 10^{-14}\right)}{\left(1 \times 10^{-2}\right)} = 1 \times 10^{-12}$$

36. **D** First let's find out how many moles of H^+ ions we need to neutralize.

Every H_2SO_3 will produce 2 H^+ ions, so for our purposes, we can think of the solution as a 0.4-molar H^+ solution.

Moles = (molarity)(volume)

Moles of H^+ = (molarity)(volume) = (0.4 M)(100 ml) = 40 millimoles = moles of OH^- required

$$\text{Volume of NaOH} = \frac{\text{moles}}{\text{molarity}} = \frac{\left(40 \text{ millimoles}\right)}{\left(0.1 \ M\right)} = 400 \text{ ml}$$

37. **C** An amphoteric species can act either as a base and gain an H^+ or as an acid and lose an H^+.

HSO_4^- can lose an H^+ to become SO_4^{2-} or it can gain an H^+ to become H_2SO_4.

HNO_3 (A), $HC_2H_3O_2$ (B), and H_3PO_4 (D) can lose only an H^+.

38. **D** Every unit of HNO_3 added to the solution will place 1 unit of H^+ ions in the solution. So first find the moles of HNO_3 added.

$$\text{Moles} = \frac{\text{grams}}{\text{MW}}$$

$$\text{Moles of } H^+ = \frac{0.630 \text{ grams}}{63.0 \text{ g/mole}} = 0.01 \text{ moles}$$

Now it's easy to find the H^+ concentration.

$$\text{Molarity} = \frac{\text{moles}}{\text{liters}}$$

$$[H^+] = \frac{0.01 \text{ moles}}{1 \text{ L}} = 0.01 \ M$$

$$pH = -\log [H^+] = -\log(0.01) = 2$$

39. **D** A buffered solution can be prepared by mixing a weak acid with an equal amount of its conjugate or by adding enough strong base to neutralize half of the weak acid present in a solution.

In (I), equal amounts of a strong acid and base are mixed. They'll neutralize each other completely to produce salt water, which is not a buffer.

In (II), enough strong base is added to neutralize half of the weak acid. This will leave equal amounts of weak acid and its conjugate base, producing a buffered solution.

In (III), equal amounts of a weak acid and its conjugate base are mixed. This will produce a buffered solution.

40. **D** The half equivalence point of this titration, shown in the plot below, is around pH = 9.8.

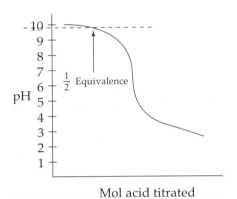

Mol acid titrated

However, this is not the pK_b of the weak base. Remember that the Henderson-Hasselbalch equation relates pH with pK_a, not pK_b. At the half equivalence point shown, the amount of base (which we can call $[A^-]$ as we do when talking about acids and pH values) is equal to the neutralized base $[HA]$. When these two quantities are equal, the H-H equation tells us that the pH = the pK_a of the acid form, which when subtracted from 14 gives the pK_b of the base. So, the pK_a of the acid form is 9.8, so the pK_b of the base is 4.2.

Constructed Response Questions

1. (a) The solubility product is the same as the equilibrium expression, but because the reactant is a solid, there is no denominator.

$$K_{sp} = [Ca^{2+}][OH^-]^2$$

(b) Use the solubility product.

$$K_{sp} = [Ca^{2+}][OH^-]^2$$

$$5.5 \times 10^{-6} = (x)(2x)^2 = 4x^3$$

$$x = 0.01 \ M \text{ for } Ca^{2+}$$

One mole of calcium hydroxide produces 1 mole of Ca^{2+}, so the concentration of $Ca(OH)_2$ must be 0.01 M.

Moles = (molarity)(volume)

Moles of $Ca(OH)_2$ = (0.01 M)(0.500 L) = 0.005 moles

Grams = (moles)(MW)

Grams of $Ca(OH)_2$ = (0.005 mol)(74 g/mol) = 0.37 g

(c)   We can find [OH⁻] from (b).

If [Ca^{2+}] = 0.01 M, then [OH⁻] must be twice that, so [OH⁻] = 0.02 M

pOH = –log[OH⁻] = 1.7

pH = 14 – pOH = 14 – 1.7 = 12.3

(d)   Find the new [OH⁻]. The hydroxide already present is small enough to ignore, so we'll use only the hydroxide just added.

$$Molarity = \frac{moles}{liters}$$

$$[OH^-] = \frac{(1.0 \text{ mol})}{(0.500 \text{ L})} = 2.0 \ M$$

Now use the K_{sp} expression.

K_{sp} = [Ca^{2+}][OH⁻]²

5.5×10^{-6} = [Ca^{2+}](2.0 M)²

[Ca^{2+}] = $1.4 \times 10^{-6} \ M$

2.   (a)   Entropy increases when a salt dissociates because aqueous particles have more randomness than a solid.

(b)   Most salt solution processes are endothermic, and endothermic processes are favored by an increase in temperature, therefore increasing temperature will increase the solubility of most salts.

(c) $Ce_2(SO_4)_3$ becomes less soluble as temperature increases, so the solution process for this salt must be exothermic.

HCl is a strong acid, which means that it dissociates completely. This means that 1 mole of HCl in solution will produce 2 moles of particles. HF is a weak acid, which means that it dissociates very little.

This means that 1 mole of HF in solution will remain at about 1 mole of particles in solution.

Therefore, the HCl solution will have more particles than the HF solution.

3. (a) Use the equilibrium expression.

$$K_1 = \frac{\left[H^+\right]\left[HCO_3^-\right]}{\left[H_2CO_3\right]}$$
$$\left[H^+\right] = \left[HCO_3^-\right] = x$$
$$\left[H_2CO_3\right] = (0.050\,M - x)$$

Assume that x is small enough that we can use $[H_2CO_3]$ = (0.050 M).

$$4.3 \times 10^{-7} = \frac{x^2}{(0.050)}$$
$$x = \left[H^+\right] = 1.5 \times 10^{-4}$$
$$pH = -\log\left[H^+\right] = -\log\left(1.5 \times 10^{-4}\right) = 3.8$$

(b) Use the equilibrium expression.

$$K_2 = \frac{\left[H^+\right]\left[CO_3^{2-}\right]}{\left[HCO_3^-\right]}$$

From (a) we know that $[H^+] = \left[HCO_3^-\right] = 1.5 \times 10^{-4}$.

$$5.6 \times 10^{-11} = \frac{\left(1.5 \times 10^{-4}\right)\left[CO_3^{2-}\right]}{\left(1.5 \times 10^{-4}\right)} = \left[CO_3^{2-}\right]$$
$$\left[CO_3^{2-}\right] = 5.6 \times 10^{-11}\,M$$

(c) (i) Adding HCl will increase [H⁺], lowering the pH.

(ii) From Le Châtelier's law, you can see that adding $NaHCO_3$ will cause the first equilibrium to shift to the left to try to use up the excess HCO_3^-. This will cause a decrease in [H⁺], raising the pH.

You may notice that adding $NaHCO_3$ will also cause the second equilibrium to shift toward the right, which should increase [H⁺], but because K_2 is much smaller than K_1, this shift is insignificant.

(iii) Adding NaOH will neutralize hydrogen ions, decreasing [H⁺] and raising the pH.

(iv) Adding NaCl will have no effect on the pH.

(d) The reaction in (d) is just the sum of the two reactions given. When two reactions can be added to give a third reaction, the equilibrium constants for those reactions can be multiplied to give K_{eq} for the third reaction.

$$K_{eq} = (K_1)(K_2) = (4.3 \times 10^{-7})(5.6 \times 10^{-11}) = 2.4 \times 10^{-17}$$

4. (a) $$K_c = \frac{[NH_3]^2}{[N_2][H_2]^3}$$

$$K_c = \frac{(0.20)^2}{(0.50)(0.20)^3} = 10$$

(b) $\Delta G = -RT \ln K$
$\Delta G = -(8.31 \text{ J/mol} \times \text{K})(570 \text{ K}) \ln(10)$
$\Delta G = (-4740)(2.3)$
$\Delta G = -11,000 \text{ J/mol}$

(c) (i) An increase in temperature favors the endothermic direction. In this case, that's the reverse reaction, so the concentration of H_2 will increase.

(ii) An increase in volume favors the direction that produces more moles of gas. In this case, that's the reverse direction, so the concentration of H_2 will increase.

(iii) According to Le Châtelier's law, increasing the concentration of the reactants forces the reaction to proceed in the direction that will use up the added reactants. In this case, adding the reactant N_2 will shift the reaction to the right and decrease the concentration of H_2.

(iv) The addition of He, a gas that takes no part in the reaction, will have no effect on the concentration of H_2.

5. (a) Use the equilibrium expression for H_2S to find $[H^+]$.

$$K_1 = \frac{\left[H^+\right]\left[HS^-\right]}{\left[H_2S\right]}$$

$[H^+] = [HS^-] = x$

$[H_2S] = 0.100\ M$

$K_1 = 1.0 \times 10^{-7}$

$$1.0 \times 10^{-7} = \frac{x^2}{\left(0.100\ M - x\right)}$$

Let's assume that 0.100 is much larger than x. That simplifies the expression.

$$1.0 \times 10^{-7} = \frac{x^2}{\left(0.100\ M\right)}$$

$x^2 = (1.0 \times 10^{-7})(0.100)\ M^2 = 1.0 \times 10^{-8}\ M^2$

$x = 1.0 \times 10^{-4}\ M = [H^+]$

$pH = -\log[H^+] = -\log(1.0 \times 10^{-4}) = 4.0$

(b) At the midpoint, $pH = pK_1$

Let's look at the Henderson-Hasselbalch expression.

$$pH = pK + \log \frac{\left[HS^-\right]}{\left[H_2S\right]}$$

When $pH = pK$, $[HS^-]$ must be equal to $[H_2S]$, making $\frac{\left[HS^-\right]}{\left[H_2S\right]}$ equal to one and $\log \frac{\left[HS^-\right]}{\left[H_2S\right]}$ equal to zero.

For every molecule of H_2S neutralized, one unit of HS^- is generated, so we must add enough NaOH to neutralize half of the HS^- initially present. (We are assuming that the further dissociation of HS^- into H^+ and S_2^- is negligible and can be ignored.)

Let's find out how many moles of H_2S were initially present.

Moles = (molarity)(volume)

Moles of H_2S = (0.100 M)(0.500 L) = 0.050 moles

We need to neutralize half of that, or 0.025 moles, so we need 0.025 moles of NaOH.

$$Volume = \frac{moles}{molarity}$$

$$Volume\ of\ NaOH = \frac{(0.025\ mol)}{(0.100\ M)} = 0.250\ L = 250\ ml$$

(c) The concentrations of the H_2S solution and NaOH solution are both 0.100 M, so the first 500 ml of NaOH solution will completely neutralize the H_2S. The final 300 ml of NaOH solution will neutralize HS^- molecules.

Initially, there were 0.050 moles of H_2S, so after 500 ml of NaOH solution was added, there were 0.050 moles of HS^-.

By adding 300 ml more of NaOH solution, we added (0.100 M) (0.300 L) = 0.030 moles of NaOH. [Moles = (molarity)(volume)]

Every unit of NaOH added neutralizes one unit of HS^-, so when all the NaOH has been added, we have 0.030 moles of S^{2-} and (0.050 moles) – (0.030 moles) = 0.020 moles of HS^-.

Now we can use the Henderson-Hasselbalch expression to find the pH. We can use the number of moles we just calculated (0.030 for S^{2-} and 0.020 for HS^-) instead of concentrations because in a solution the concentrations are proportional to the number of moles.

$$pH = pK + \log \frac{\left[S^{2-} \right]}{\left[HS^- \right]}$$
$$pK = -\log(1.3 \times 10^{-13}) = 12.9$$
$$\log \frac{\left[S^{2-} \right]}{\left[HS^- \right]} = \log \frac{(0.030)}{(0.020)} = \log(1.5) = 0.18$$

$$pH = 12.9 + 0.18 = 13.1.$$

(d) The reaction ($H_2S \rightleftharpoons 2\ H^+ + S^{2-}$) is the sum of the two acid dissociation reactions below.

$$H_2S \rightleftharpoons H^+ + HS^-$$

$$HS^- \rightleftharpoons H^+ + S^{2-}$$

If one reaction is the sum of two other reactions, then its equilibrium constant will be the product of the equilibrium constants for the other two reactions.

So $K_{eq} = K_1K_2 = (1.0 \times 10^{-7})(1.3 \times 10^{-13}) = 1.3 \times 10^{-20}$

6. (a) $pK_b = -\log K_b = -\log(1.79 \times 10^{-5}) = 4.75$

(b) This is a buffered solution, so we'll use the Henderson-Hasselbalch expression.

First let's find $[NH_4^+]$ and $[NH_3]$.

Moles = (molarity)(volume)

Moles of $NH_4^+ = (0.100\ M)(0.100\ L) = 0.010$ moles

Moles of $NH_3 = (0.200)(0.080\ L) = 0.016$ moles

When we mix the solutions, the volume becomes (0.100 L) + (0.080 L) = 0.180 L.

$$\text{Molarity} = \frac{\text{moles}}{\text{volume}}$$

$[NH_4^+] = \dfrac{(0.010\ \text{mol})}{(0.180\ \text{L})} = 0.056\ M$

$[NH_3] = \dfrac{(0.016\ \text{mol})}{(0.180\ \text{mol})} = 0.089\ M$

Now we can use the Henderson-Hasselbalch expression for bases.

$$pOH = pK + \log\frac{[NH_4^+]}{[NH_3]}$$

$pK = 4.75$

$\log\dfrac{[NH_4^+]}{[NH_3]} = \log\dfrac{(0.056\ M)}{(0.089\ M)} = -0.20$

$pOH = 4.75 + (-0.20) = 4.55$

$pH = 14 - pOH = 14 - 4.55 = 9.45$

(c) First let's find out how many moles of NaOH were added.

$$\text{Moles} = \frac{\text{moles}}{\text{MW}}$$

Moles of $NaOH = \dfrac{(0.200\ \text{g})}{(40.0\ \text{g/m})} = 0.005\ \text{mol}$

When NaOH is added to the solution, the following reaction occurs:

$$NH_4^+ + OH^- \rightarrow NH_3 + H_2O$$

So for every unit of NaOH added, one ion of NH_4^+ disappears and one molecule of NH_3 appears. We can use the results of the molar calculations we did in part (b).

Moles of NH_4^+ = (0.010) – (0.005) = 0.005 moles

Moles of NH_3 = (0.016) + (0.005) = 0.021 moles

Now we can use the Henderson-Hasselbalch expression. We can use the number of moles we just calculated (0.005 for NH_4^+ and 0.021 for NH_3) instead of concentrations because in a solution the concentrations will be proportional to the number of moles.

$$pOH = pK + \log \frac{[NH_4^+]}{[NH_3]}$$

$$pK = 4.75$$

$$\log \frac{[NH_4^+]}{[NH_3]} = \log \frac{(0.005 \text{ mol})}{(0.021 \text{ mol})} = -0.62$$

$$pOH = 4.75 + (-0.62) = 4.13$$

$$pH = 14 - pOH = 14 - 4.13 = 9.87$$

(d)   When equal quantities of a base (NH_3) and its conjugate acid (NH_4^+) are mixed in a solution, the pOH will be equal to the pK_b.

From (a), $pOH = pK_b = 4.75$

$pH = 14 - pOH = 14 - 4.75 = 9.25$.

7.   (a)   The hypochlorous acid will donate its proton to the hydroxide. The sodium ions are spectators and would not appear in the net ionic equation.

$HOCl + OH^- \leftrightarrow OCl^- + H_2O$ (*l*)

(b)   At the equivalence point, the moles of acid is equal to the moles of base.

Moles of base = (1.47 M)(0.03423 L) = 0.0503 mol

Concentration acid = (0.0503 mol/0.02500 L) = 2.01 M

(c)   This calls for an ICE chart. We will first need the moles of both the acid and base.

Moles HOCl = (2.01 M)(0.02500 L) = 0.0503 mol HOCl

Moles OH^- = (1.47 M)(0.02855 L) = 0.0420 mol OH^-

Putting those numbers into our ICE chart:

	HOCl +	OH⁻ ↔	OCl⁻ +	H₂O
Initial	0.0503	0.0420	0	X
Change	−0.0420	−0.0420	0.0420	X
Equilibrium	0.0083	0	0.0420	X

Now that we know the number of moles at equilibrium of each species, we need to determine their new concentrations. The total volume of the solution at equilibrium is 25.00 mL + 28.55 mL = 53.55 mL. So the concentrations at equilibrium will be:

$[HOCl] = 0.0083$ mol/0.05355 L = 0.15 M
$[OCl^-]$ 0.0420 mol/0.05355 L = 0.784 M

To finish off we use Henderson-Hasselbalch:

$$pH = pK_a + \log \frac{[OCl^-]}{[HOCl]}$$

$$pH = -\log(3.5 \times 10^{-8}) + \log \frac{(0.784)}{(0.15)}$$

$$pH = 7.5 + \log(5.2)$$

$$pH = 7.5 + 0.72$$

$$pH = 8.2$$

(d) (i) If additional NaOH is added, that would mean a larger number of moles of NaOH added to the flask, and thus more apparent moles of HOCl would be present at the equivalence point. More apparent moles of HOCl would lead to a larger numerator in the acid molarity calculation and thus a larger apparent molarity. As the student's calculated molarity was too low, this could not have caused the error.

(ii) If the buret was not rinsed with the NaOH prior to filling it, the concentration of the NaOH would be diluted by the water inside the buret. The student would then have to add a greater volume of NaOH to reach equivalence. However, in the calculations using the original concentration of NaOH multiplied by the higher volume would lead to an artificially high number of moles of NaOH and thus, more apparent moles HOCl at equivalence. More apparent moles of HOCl would lead to a larger numerator in the acid molarity calculation and thus a larger apparent molarity. As the student's calculated molarity was too low, this could not have caused the error.

(iii) If the student did not add enough acid to the flask, that would cause the denominator in the molarity of the acid calculation to be artificially high. This, in turn, would make the calculated acid molarity be artificially low. This matches up with the student's results and could be an acceptable source of error.

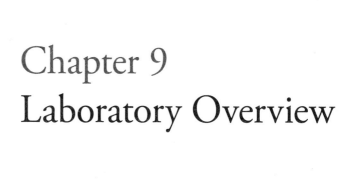

Chapter 9
Laboratory Overview

INTRODUCTION

The AP Chemistry exam will test your knowledge of basic lab techniques, as well your understanding of accuracy and precision and your ability to analyze potential sources of error in a lab. In this section, we will discuss safety and accuracy precautions, laboratory equipment, and laboratory procedures.

SAFETY

Here are some basic safety rules that might turn up in test questions.

- Don't put chemicals in your mouth. You were told this when you were four years old, and it still holds true for the AP Chemistry Exam.
- When diluting an acid, always add the acid to the water. This is to avoid the spattering of hot solution.
- Always work with good ventilation; many common chemicals are toxic.
- When heating substances, do it slowly. When you heat things too quickly, they can spatter, burn, or explode.

ACCURACY

Here are some rules for ensuring the accuracy of experimental results.

- When titrating, rinse the buret with the solution to be used in the titration instead of with water. If you rinse the buret with water, you might dilute the solution, which will cause the volume added from the buret to be too large.
- Allow hot objects to return to room temperature before weighing. Hot objects on a scale create convection currents that may make the object seem lighter than it is.
- Don't weigh reagents directly on a scale. Use a glass or porcelain zcon- tainer to prevent corrosion of the balance pan.
- When collecting a gas over water, remember to take into account the pressure and volume of the water vapor.
- Don't contaminate your chemicals. Never insert another piece of equipment into a bottle containing a chemical. Instead you should always pour the chemical into another clean container. Also, don't let the inside of the stopper for a bottle containing a chemical touch another surface.
- When mixing chemicals, stir slowly to ensure even distribution.
- Be conscious of significant figures when you record your results. The number of significant figures you use should indicate the accuracy of your results.
- Be aware of the difference between accuracy and precision. A measurement is accurate if it is close to the accepted value. A series of measurements is precise if the values of all of the measurements are close together.

SIGNIFICANT FIGURES

When taking measurements in lab, your measurement will always have a certain number of significant figures. For instance, is using a balance to mass something to the hundredths place, you might get 23.15 g. That number has four significant figures. The balance is no more accurate than that; you could not say the mass of the object is 23.15224 g.

It's important to be able to identify the number of significant figures in any number given to you on the AP exam. There's a plethora of information on ways to count significant figures out there, but we've been able to reduce it to two simple rules:

1. For numbers without a decimal place, you count every number except for trailing zeroes (those which appear after all non-zeroes). So, 100 mL only has one sig fig, and 250 mL has two. On the other hand, 105 g has three significant figures- the zero in that measurement does not trail all other numbers.
2. For numbers with a decimal place, you count every number except for leading zeroes (those which appear before all non-zeroes). A number like 0.052 has 2 significant figures, but 0.0520 g would have three (trailing zeroes DO count in numbers with decimals). Most (but not all) values you get on the AP exam will have decimal places.

This leads to numbers that otherwise might look very strange. For instance, it is not unusual to see numbers with a decimal at the end, but no numbers past it. That's because a number like 100. g has three significant figures, but 100 g only has one. In science, 100. $g \neq 100$ g $\neq 100.0$ g. All of those values have different numbers of significant figures, and that implies different levels of accuracy. significant figures only apply to measurements with units; pure unitless numbers (such as those you find in math) do not follow these rules. Nonetheless, many math books do have units on practice problems and then ignore significant figures. Yes, it's wrong to do that. But, don't tell your math teacher- they typically dislike significant figures.

When doing calculations, it is important that any calculated value cannot be more accurate than the measurements used in the calculation. Essentially, significant figures can tell you how to round your answers correctly. Again, this can be divided into two categories:

1. When multiplying and dividing, your answer cannot have more significant figures than your least accurate measurement.

    For instance:

    2.50 g / 12 cm^3 = 0.20833 g/cm^3

    is wrong. Your two measurements have three significant figures and two significant figures, respectively. Your answer cannot have five. It has to have only two; because that's how many your least accurate measurement had. The correct answer is 0.21 g/cm^3

2. When adding and subtracting, your answer cannot have more figures after the decimal place than your value with the least number of figures after its decimal place.

For instance:

1.435 cm + 12.1 cm = 13.535 cm

is wrong. Your values have three figures after the decimal and one figure after the decimal, respectively. Your answer cannot have more than one figure past the decimal because that's how many your least accurate measurement had. The correct answer is 13.5 cm.

On a side note, counting numbers are considered to have an infinite number of significant figures. If you are doing a molar mass conversion and you identify the conversion as 22.99 g of sodium = 1 mol of sodium, your eventual answer will have four significant figures. That 1 mol represents *exactly* 1 mol of sodium, essentially, a 1 with an infinite number of zeroes past the decimal place. This will usually come up when dealing with stoichiometry—the coefficients in a balanced equation are counting numbers of moles, and don't count when considering the number of significant figures your answer should have.

It's a good habit to get into making sure your calculations always have the correct number of significant figures as you do them, both in lab and on any practice problems you do in class. The AP exam may have questions on significant figures in lab; they are typically integrated into the constructed response section. Additionally, you may lose points for the incorrect number of significant figures when you do calculations on constructed response section. Along with making sure you have the correct units on your number, making sure you do your calculations to the correct number of significant figures is good laboratory practice."

METHODS OF SEPARATION

Filtration—In filtration, solids are separated from liquids when the mixture is passed through a filter. Typically, porous paper is used as the filter. To find the amount of solid that is filtered out of a mixture, the filter paper containing the solid is allowed to dry and then weighed. The initial weight of the clean, dry filter is then subtracted from the weight of the dried filter paper and solid.

Distillation—In distillation, the differences in the boiling points of liquids are used to separate them. The temperature of the mixture is raised to a temperature that is greater than the boiling point of the more volatile substance and lower than the boiling point of the less volatile substance. The more volatile substance will vaporize, leaving the less volatile substance.

Chromatography—In chromatography, substances are separated by the differences in the degree to which they are adsorbed onto a surface. The substances are passed over the adsorbing surface, and the ones that stick to the surface with greater attraction will move slower than the substances that are less attracted to the surface. This difference in speeds separates the substances. The name chromatography came about because the process is used to separate pigments.

MEASURING CONCENTRATION

Titration is one of the most important laboratory procedures. In titration, an acid-base neutralization reaction is used to find the concentration of an unknown acid or base. It takes exactly one mole of hydroxide ions (base) to neutralize one mole of hydrogen ions (acid), so the concentration of an unknown acid solution found by calculating how much of a known basic solution is required to neutralize a sample of given volume. The most important formula in titration experiments is derived from the definition of molarity.

$$\text{Molarity} = \text{Moles/Liters}$$
$$\text{Moles} = (\text{Molarity})(\text{Liters})$$

The moment when exactly enough base has been added to the sample to neutralize the acid present is called the equivalence point. In the lab, an indicator is used to tell when the equivalence point has been reached. An indicator is a substance that is one color in acid solution and a different color in basic solution. Two popular indicators are phenolphthalein, which is clear in acidic solution and pink in basic solution, and litmus, which is pink in acidic solution and blue in basic solution.

Spectrophotometer—A spectrophotometer measures slight variations in color. It can be used to measure the concentration of ions that produce colored solutions.

IDENTIFYING CHEMICALS IN SOLUTION

Precipitation

One of the most useful ways of identifying unknown ions in solution is precipitation. You can use the solubility rules given in the solubility chapter to see how the addition of certain ions to solution will cause the specific precipitation of other ions. For instance, the fact that $BaSO_4$ is insoluble can be used to identify either can be used to identify either Ba^{2+} or SO_4^{2-} in solution; if the solution contains Ba^{2+} ions, then the addition of SO_4^{2-} will cause a precipitation reaction. The inverse is true for a solution that contains SO_4^{2-} ions.

In the same way, the insolubility of $AgCl$ can be used to identify either Ag^+ or Cl^- in solution.

Conduction

You can tell whether a solution contains ions by checking to see if the solution conducts electricity. Ionic solutes conduct electricity in solution; nonionic solutes do not.

Flame Tests

Some ions burn with distinctly colored flames. Flame tests can be used to identify Li^+ (red), Na^+ (yellow), and K^+ (purple), as well as the other alkali metals. The alkaline earths, including Ba^{2+} (green), Sr^{2+} (red), and Ca^{2+} (red) also burn with colored flames.

Acid-Base Reaction

- When a base is added to an NH_4^+ solution, the distinctive odor of ammonia can be detected.
- When an acid is added to a solution containing S_2^-, the rotten-egg odor of H_2S can be detected.
- When acid is added to a solution containing CO_3, CO_2 gas is produced.

LABORATORY EQUIPMENT

The pictures below show some standard chemistry lab equipment.

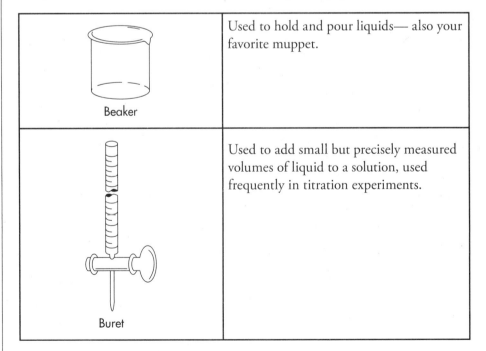

Beaker	Used to hold and pour liquids— also your favorite muppet.
Buret	Used to add small but precisely measured volumes of liquid to a solution, used frequently in titration experiments.

Burner	Used to apply heat and to wake up sleeping AP Chem students.
Crucible tongs	Used to handle objects that are too hot to touch (careful though, they can break test tubes!).
Dropper pipette	Used to add small amounts of liquid to a solution, but only when a precise volume is not needed because the drops themselves should be consistent only for one particular dropper (i.e. adding an indicator, comparing the amount of drops of strong acid or base needed for a pH change in different solutions).
Erlenmeyer flask	A flask used for heating liquids. The conic shape allows stirring.
Evaporating dish	Used to hold liquids for evaporation. The wide mouth allows vapor to escape.
Florence flask	Used for boiling of liquids. The small neck prevents excessive evaporation and splashing.

Forceps	A fancy name for tweezers.
Funnel	Used to get liquids into a smaller container. Doubles as a cute hat when inverted.
Graduated cylinder	Used for measuring precisely a volume of liquid to be poured all at once (rather than dripped from a buret).
Graduated pipette	Used to transfer small and precise volumes of liquid from one container to another (gradations indicate the volume).
Metal spatula	Used to scoop and transport powders.

Mortar and pestle	Used to grind solids into powders suitable for dissolving or mixing.
Pipette bulb	Rubber bulb used to draw liquid into pipette. In a pinch could be used as a miniature clown nose.
Platform balance (triple beam)	A very precise scale operated by moving a set of three weights (typically corresponding to 100, 10, and 1 gram increments). A measurement will proceed like this: Rear weight is in the notch reading 30 g Middle weight is in the notch reading 200 g Front beam weight reads 3.86 g The sample weighs 200 + 30 + 3.86 = 233.86 g
Ring clamp	Used to hold funnels or other vessels in conjunction with a stand.
Rubber policeman	A hard tipped rubber scraper used to transfer precipitate.

Safety goggles	Used by chemists worldwide to protect their eyes during all laboratory experiments.
Test tube	Used to contain samples, especially when heating. Sometimes babies come out of them.
Thermometer	Measures temperature of a solution (don't use these to measure your body temperature, you don't know where they've been).
Volumetric pipette	Used to transfer small amounts of liquid. The big difference between a graduated pipette and a volumetric pipette is that the volumetric type is suited for only one particular volume. Because of this, they are extremely accurate.

CHAPTER 9 QUESTIONS

Multiple-Choice Questions

Questions 1–4

    (A) Oxidation-reduction
    (B) Neutralization
    (C) Fusion
    (D) Combination

    Which of the reaction types listed above best describes each of these processes?

1. $CO_2(g) + CaO(s) \rightarrow CaCO_3(s)$

2. $2\ Fe^{3+}(aq) + 2\ I^-(aq) \rightarrow 2\ Fe^{2+}(aq) + I_2(aq)$

3. $CH_3COOH(aq) + NaOH(aq) \rightarrow CH_3COONa(aq) + H_2O(l)$

4. $CH_4(g) + 2\ O_2(g) \rightarrow CO_2(g) + 2\ H_2O(g)$

Questions 5–7

    (A) Na^+
    (B) Cu^{2+}
    (C) Ag^+
    (D) Al^{3+}

5. This ion turns an aqueous solution deep blue.

6. This ion forms a white precipitate when added to a solution containing chloride ions.

7. This ion produces a yellow flame when burned.

8. Which of the following indicators would be most useful in identifying the equivalence point of a titration for a solution that has a hydrogen ion concentration of $7 \times 10^{-4}\ M$ at the equivalence point?

    (A) Methyl violet (pH range for color change is 0.1–2.0)
    (B) Methyl yellow (pH range for color change is 1.2–2.3)
    (C) Methyl orange (pH range for color change is 2.9–4.0)
    (D) Methyl red (pH range for color change is 4.3–6.2)

9. The volume of a liquid is to be measured. Which of the following cylindrical flasks would take the most accurate measurement?

    (A) A flask with 1 ml gradations and a diameter of 1 cm
    (B) A flask with 1 ml gradations and a diameter of 3 cm
    (C) A flask with 5 ml gradations and a diameter of 1 cm
    (D) A flask with 5 ml gradations and a diameter of 3 cm

10. Which of the following is (are) considered to be proper laboratory procedure?

    I. Reading the height of a fluid in a buret from a point level with the fluid's meniscus.
    II. Placing a sample to be weighed directly on the pan of the balance.
    III. Stirring a solution constantly during a titration.

(A) I only
(B) III only
(C) I and III only
(D) I, II, and III

11. A 0.1-molar NaOH solution is to be released from a buret in a titration experiment to measure the hydrogen ion concentration of an unknown acid. Which of the following laboratory procedures would cause an error in the measure of the concentration of the acid?

    I. The buret was rinsed with the NaOH solution immediately before the titration.
    II. The buret was rinsed with distilled water immediately before the titration.
    III. The buret was rinsed with the unknown acid immediately before the titration.

(A) I only
(B) III only
(C) I and II only
(D) II and III only

12. Which of the following is NOT proper procedure for transferring a solution with a pipette?

(A) Rinsing the pipette with the solution to be transferred.
(B) Using your mouth to draw the solution into the pipette.
(C) Covering the top of the pipette with your index finger to keep the solution from escaping.
(D) Draining the solution into a waste beaker until the meniscus drops to the calibration mark.

13. An object that was weighed on a balance was later found to be slightly heavier than the weight that was recorded by the balance. Which of the following could have caused the discrepancy?

    I. There was some foreign matter on the weighing paper along with the object.
    II. The object was hot when it was weighed.
    III. The experimenter neglected to account for the weight of the weighing paper.

(A) I only
(B) II only
(C) I and II only
(D) I, II, and III

14. An experimenter wishes to use test paper to find the pH of a solution. Which of the following is part of the proper procedure for this process?

(A) Dipping the test paper in the solution while stirring.
(B) Dipping the test paper in the solution without stirring.
(C) Pouring some of the solution onto the dry test paper.
(D) Dipping the test paper in distilled water and slowly adding the solution to the water while stirring.

Constructed Response Questions

1. An experiment was conducted to determine the molecular weight of a pure salt sample. The mass of the salt sample was known. The salt was dissolved in a container of water of known mass and the freezing point of the solution was measured. The molecular weight was calculated by the freezing point depression method. How would the calculated value of the molecular weight be affected by each of the following?

   (a) The experimenter failed to take the dissociation of the salt into account.
   (b) The experimenter mistook molarity for molality, and used liters of solution instead of kilograms of solvent in the calculation to find the number of moles of solute.
   (c) The container used for the experiment was not rinsed and contained dust particles.
   (d) The experimenter misread the thermometer and recorded a freezing point that was higher than the true value.
   (e) The experimenter did not notice that some solid salt did not completely dissolve.

2. Use your knowledge of chemical principles to answer or explain each of the following:

   (a) When helium gas is to be collected in a jar by the displacement of air, the opening of the jar must be directed downward. When carbon dioxide gas is to be collected in a jar by the displacement of air, the opening of the jar must be directed upward.
   (b) Will the molar quantity calculated for a gas collected over water be too large or too small if the experimenter fails to take into account the vapor pressure of water?
   (c) Why is it easier to separate oxygen gas from hydrogen gas by the method of successive effusion than it is to separate oxygen gas from nitrogen gas by the same method?
   (d) Give an explanation for why an attempt to separate two liquids by distillation may fail.

CHAPTER 9 ANSWERS AND EXPLANATIONS

Multiple-Choice Questions

1. **D** In a combination (or composition, or synthesis) reaction, two substances combine to form a more complex substance.

    A combination reaction is the opposite of a decomposition reaction, so if this reaction occurred in reverse, it would be a decomposition reaction.

2. **A** In an oxidation-reduction reaction, electrons are transferred between the reactants, causing the oxidation state of the element that is oxidized to increase and the oxidation state of the element that is reduced to decrease. In this reaction, Fe^{3+} is reduced to Fe^{2+} and I^- is oxidized to I^0.

3. **B** In a neutralization reaction, an acid (in this case, CH_3COOH) and a base (NaOH) react to form water and a salt (CH_3COONa).

4. **A** This reaction, the combustion of an organic compound, is also an oxidation-reduction reaction. As we said above, in a redox reaction, electrons are transferred between the reactants, causing the oxidation state of the element that is oxidized to increase and the oxidation state of the element that is reduced to decrease. In this reaction, H^- is oxidized to H^+ and O^0 is reduced to O^{2-}.

5. **B** Copper, along with most of the transition metals, forms colored solutions with water. This is true because the d electrons of the transition metals are constantly changing energy levels and emitting radiation in the visible spectrum.

6. **C** Silver is the only ion listed that forms an insoluble chloride.

7. **A** Sodium, along with the other Group IA elements, produces a colored flame in the flame test.

8. **C** If $[H^+]$ is 7×10^{-4} M, then the pH must be between 3 and 4. Only methyl orange changes color between 3 and 4.

9. **A** Volume = (height)(cross-sectional area)

    The smaller the gradations, the more accurately the height can be measured. The smaller the area, the farther apart the 1 ml gradations will be and the more accurately the height of the fluid can be measured.

10. C   Choices (I) and (III) are proper experimental procedures.

(II) is not; a sample should always be weighed in a glass or porcelain container to prevent a reaction with the balance pan.

11. D   Rinsing the buret with the NaOH solution (I) is proper procedure and will not change the concentration of the NaOH solution and will not cause an error. Rinsing the buret with distilled water (II) will dilute the NaOH solution, lowering the concentration and causing an error, and rinsing the buret with the unknown acid (III) will cause some of the NaOH solution to be neutralized in the buret, lowering its concentration and causing an error.

12. B   You should never use your mouth to draw solution into a pipette. Instead, you should use a rubber suction bulb.

All of the other choices are part of the proper procedure.

13. B   When a hot object is weighed (II), convection currents around the object can reduce the apparent mass measured by the balance.

Choices (I) and (III) would both cause the weight measured by the balance to be greater than the actual weight of the object. We're looking for the opposite effect.

14. C   (A) and (B) are wrong because there is a danger of contaminating the solution by adding the paper.

(D) is wrong because adding the solution to distilled water completely changes the solution and defeats the purpose of testing it.

So pouring the solution onto the dry test paper (C) is the proper procedure.

Constructed Response Questions

1. A note for the answers:

    In this experiment, the freezing point of the solution is measured, and from the freezing point the freezing-point depression, ΔT, is calculated.

    The equation $m = \dfrac{\Delta T}{kx}$ is used to calculate the molality of the solution.

    Moles = (molality)(kg of solvent) is used to calculate the number of moles of salt.

    $MW = \dfrac{moles}{grams}$ is used to calculate the molecular weight of the salt.

    (a) The experimenter makes $x = 1$, instead of 2 or 3. So the calculated value of m will be too large, the calculated value of moles of salt will be too large, and the calculated MW will be too small.

    (b) Moles = (molality)(kg of solvent)
        Moles = (molarity)(liters of solution)

    Because the solvent is water, the distinction between kilograms and liters is not important (for water, 1kg = 1 L), but the distinction between solvent and solution may make a difference.

    Liters of solution will be a little larger than kilograms of solvent, so the calculated value for moles of salt will be too large, and the calculated MW will be too small.

    (c) Extra particles in the solution will cause the measured freezing point to be too low. ΔT will be too large, m will be too large, and the calculated MW will be too small.

    (d) If the freezing point is too high, the calculated ΔT will be too small, m will be too small, and the calculated MW will be too large.

    (e) If some salt does not dissolve, then the grams of salt used in the calculation will be larger than the amount actually in the solution, and the calculated MW will be too large.

2. (a) Helium (MW = 4 g/mol) is less dense than air, so it will rise to the top of the jar, displacing air downward.

Carbon dioxide (MW = 44 g/mol) is denser than air, so it will sink to the bottom of the jar, displacing air upward.

(b) If the vapor pressure from water is ignored, the pressure of the gas used in the calculation will be too large.

$$n = \frac{PV}{RT}$$

If P is too large, then n, the calculated molar quantity of gas, will be too large.

(c) Separation of gases by successive effusion depends on Graham's law.

$$\frac{v_1}{v_2} = \sqrt{\frac{MW_2}{MW_1}}$$

The greater the difference in molecular weights, the greater the difference in average molecular speeds, and the greater the difference in rates of effusion.

Oxygen gas (32 g/mol) and nitrogen gas (28 g/mol) have similar molecular weights.

Oxygen gas (32 g/mol) and hydrogen gas (2 g/mol) have very different molecular weights.

(d) The most likely reason for the failure of separation by distillation would be that the boiling points of the two liquids are too close together.

Part V
The Princeton Review AP Chemistry Practice Tests

Chapter 10
Practice Test 1

AP® Chemistry Exam

SECTION I: Multiple-Choice Questions

DO NOT OPEN THIS BOOKLET UNTIL YOU ARE TOLD TO DO SO.

At a Glance

Total Time
1 hour and 30 minutes
Number of Questions
60
Percent of Total Grade
50%
Writing Instrument
Pencil required

Instructions

Section I of this examination contains 60 multiple-choice questions. Fill in only the ovals for numbers 1 through 60 on your answer sheet.

CALCULATORS MAY NOT BE USED IN THIS PART OF THE EXAMINATION.

Indicate all of your answers to the multiple-choice questions on the answer sheet. No credit will be given for anything written in this exam booklet, but you may use the booklet for notes or scratch work. After you have decided which of the suggested answers is best, completely fill in the corresponding oval on the answer sheet. Give only one answer to each question. If you change an answer, be sure that the previous mark is erased completely. Here is a sample question and answer.

Sample Question   Sample Answer

Chicago is a
(A) state
(B) city
(C) country
(D) continent

Use your time effectively, working as quickly as you can without losing accuracy. Do not spend too much time on any one question. Go on to other questions and come back to the ones you have not answered if you have time. It is not expected that everyone will know the answers to all the multiple-choice questions.

About Guessing

Many candidates wonder whether or not to guess the answers to questions about which they are not certain. Multiple choice scores are based on the number of questions answered correctly. Points are not deducted for incorrect answers, and no points are awarded for unanswered questions. Because points are not deducted for incorrect answers, you are encouraged to answer all multiple-choice questions. On any questions you do not know the answer to, you should eliminate as many choices as you can, and then select the best answer among the remaining choices.

Disclaimer

This test is an approximation of the test that you will take. For up-to-date information, please remember to check the AP Central website.

CHEMISTRY
SECTION I
Time—1 hour and 30 minutes

Material in the following table may be useful in answering the questions in this section of the examination.

PERIODIC CHART OF THE ELEMENTS

1 H 1.0																	2 He 4.0
3 Li 6.9	4 Be 9.0											5 B 10.8	6 C 12.0	7 N 14.0	8 O 16.0	9 F 19.0	10 Ne 20.2
11 Na 23.0	12 Mg 24.3											13 Al 27.0	14 Si 28.1	15 P 31.0	16 S 32.1	17 Cl 35.5	18 Ar 39.9
19 K 39.1	20 Ca 40.1	21 Sc 45.0	22 Ti 47.9	23 V 50.9	24 Cr 52.0	25 Mn 54.9	26 Fe 55.8	27 Co 58.9	28 Ni 58.7	29 Cu 63.5	30 Zn 65.4	31 Ga 69.7	32 Ge 72.6	33 As 74.9	34 Se 79.0	35 Br 79.9	36 Kr 83.8
37 Rb 85.5	38 Sr 87.6	39 Y 88.9	40 Zr 91.2	41 Nb 92.9	42 Mo 95.9	43 Tc (98)	44 Ru 101.1	45 Rh 102.9	46 Pd 106.4	47 Ag 107.9	48 Cd 112.4	49 In 114.8	50 Sn 118.7	51 Sb 121.8	52 Te 127.6	53 I 126.9	54 Xe 131.3
55 Cs 132.9	56 Ba 137.3	57 *La 138.9	72 Hf 178.5	73 Ta 180.9	74 W 183.9	75 Re 186.2	76 Os 190.2	77 Ir 192.2	78 Pt 195.1	79 Au 197.0	80 Hg 200.6	81 Tl 204.4	82 Pb 207.2	83 Bi 209.0	84 Po (209)	85 At (210)	86 Rn (222)
87 Fr (223)	88 Ra 226.0	89 †Ac 227.0															

*Lanthanum Series

58 Ce 140.1	59 Pr 140.9	60 Nd 144.2	61 Pm (145)	62 Sm 150.4	63 Eu 152.0	64 Gd 157.3	65 Tb 158.9	66 Dy 162.5	67 Ho 164.9	68 Er 167.3	69 Tm 168.9	70 Yb 173.0	71 Lu 175.0

†Actinium Series

90 Th 232.0	91 Pa 231.0	92 U 238.0	93 Np 237.0	94 Pu (244)	95 Am (243)	96 Cm (247)	97 Bk (247)	98 Cf (251)	99 Es (252)	100 Fm (258)	101 Md (258)	102 No (259)	103 Lr (260)

DO NOT DETACH FROM BOOK.

GO ON TO THE NEXT PAGE.

1. $2\ ClF\ (g) + O_2\ (g) \leftrightarrow Cl_2O\ (g) + F_2O\ (g)\quad \Delta H = 167\ kJ$

    During the reaction above, the product yield can be increased by increasing the temperature of the reaction. Why is this effective?

    (A) The reaction is endothermic; therefore adding heat will shift it to the right.
    (B) Increasing the temperature increases the speed of the molecules, meaning there will be more collisions between them.
    (C) The reactants are less massive than the products, and an increase in temperature will cause their kinetic energy to increase more than that of the products.
    (D) The increase in temperature allows for a higher percentage of molecular collisions to occur with the proper orientation to create the product.

2. $CH_3NH_2\ (aq) + H_2O\ (l) \leftrightarrow OH^-\ (aq) + CH_3NH_3^+\ (aq)$

    The above reaction represents the reaction between the base methylamine ($K_b = 4.38 \times 10^{-4}$) and water. Which of the following best represents the concentrations of the various species at equilibrium?

    (A) $[OH^-] > [CH_3NH_2] = [CH_3NH_3^+]$
    (B) $[OH^-] = [CH_3NH_2] = [CH_3NH_3^+]$
    (C) $[CH_3NH_2] > [OH^-] > [CH_3NH_3^+]$
    (D) $[CH_3NH_2] > [OH^-] = [CH_3NH_3^+]$

3. The diagram below shows the relative atomic sizes of three different elements from the same period. What of the following must be true about the elements?

    (A) The effective nuclear charge would be the greatest in element X.
    (B) The first ionization energy will be greatest in element X.
    (C) The electron shielding effect will be greatest in element Z.
    (D) The electronegativity value will be greatest in element Z.

4. A sealed, rigid container contains three gases: 28.0 g of nitrogen, 40.0 g of argon, and 36.0 g of water vapor. If the total pressure exerted by the gases is 2.0 atm, what is the partial pressure of the nitrogen?

    (A) 0.33 atm
    (B) 0.40 atm
    (C) 0.50 atm
    (D) 2.0 atm

5. A sample of liquid NH_3 is brought to its boiling point. Which of the following occurs during the boiling process?

    (A) The N-H bonds within the NH_3 molecules together break apart.
    (B) The overall temperature of the solution rises as the NH_3 molecules speed up.
    (C) The amount of energy within the system remains constant.
    (D) The hydrogen bonds holding separate NH_3 molecules together break apart.

GO ON TO THE NEXT PAGE.

Questions 6-10 refer to the following.

Two half-cells are set up as follows:

Half-Cell A: Strip of Cu (s) in CuNO₃ (aq)
Half-Cell B: Strip of Zn (s) in Zn(NO₃)₂ (aq)

When the cells are connected according to the diagram below, the following reaction occurs:

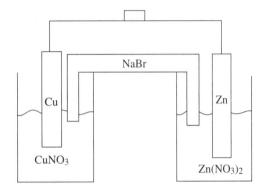

$$2\,Cu^+\,(aq) + Zn\,(s) \rightarrow 2\,Cu\,(s) + Zn^{2+}\,(aq)\quad E° = +1.28\,V$$

6. Correctly identify the anode and cathode in this reaction as well as where oxidation and reduction are taking place.

   (A) Cu is the anode where oxidation occurs, and Zn is the cathode where reduction occurs.
   (B) Cu is the anode where reduction occurs, and Zn is the cathode where oxidation occurs.
   (C) Zn is the anode where oxidation occurs, and Cu is the cathode where reduction occurs.
   (D) Zn is the anode where reduction occurs, and Cu is the cathode where oxidation occurs.

7. How many moles of electrons must be transferred to create 127 g of copper?

   (A) 1 mole of electrons
   (B) 2 moles of electrons
   (C) 3 moles of electrons
   (D) 4 moles of electrons

8. If the $Cu^+ + e^- \rightarrow Cu\,(s)$ half reaction has a standard reduction potential of +0.52 V, what is the standard reduction potential for the $Zn^{2+} + 2e^- \rightarrow Zn\,(s)$ half reaction?

   (A) +0.76 V
   (B) −0.76 V
   (C) +0.24 V
   (D) −0.24 V

9. As the reaction progresses, what will happen to the overall voltage of the cell?

   (A) It will increase as [Zn^{2+}] increases.
   (B) It will increase as [Cu^+] increases.
   (C) It will decrease as [Zn^{2+}] increases.
   (D) The voltage will remain constant.

10. What will happen in the salt bridge as the reaction progresses?

   (A) The Na^+ ions will flow to the Cu/Cu⁺ half-cell.
   (B) The Br⁻ ions will flow to the Cu/Cu⁺ half-cell.
   (C) Electrons will transfer from the Cu/Cu⁺ half-cell to the Zn/Zn²⁺ half-cell.
   (D) Electrons will transfer from the Zn/Zn²⁺ half-cell to the Cu/Cu⁺ half-cell.

---

11. For a reaction involving nitrogen monoxide inside a sealed flask, the value for the reaction quotient (Q) was found to be 1.1×10^2 at a given point. If, after this point, the amount of NO gas in the flask increased, which reaction is most likely taking place in the flask?

   (A) $NOBr\,(g) \leftrightarrow NO\,(g) + \tfrac{1}{2}\,Br_2\,(g)$ $K_c = 3.4 \times 10^{-2}$
   (B) $2\,NOCl\,(g) \leftrightarrow 2\,NO\,(g) + Cl_2\,(g)$ $K_c = 1.6 \times 10^{-5}$
   (C) $2\,NO\,(g) + 2\,H_2\,(g) \leftrightarrow N_2\,(g) + 2\,H_2O\,(g)$
        $K_c = 4.0 \times 10^6$
   (D) $N_2\,(g) + O_2\,(g) \leftrightarrow 2\,NO\,(g)$ $K_c = 4.2 \times 10^2$

12. Which of the following substances has an asymmetrical molecular structure?

   (A) SF_4
   (B) PCl_5
   (C) BF_3
   (D) CO_2

GO ON TO THE NEXT PAGE.

13.

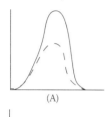

The diagram above shows the speed distribution of molecules in a gas held at 200 K. Which of the following representations would best represent the gas at a higher temperature? (Note: The original line is shown as a dashed line in the answer options.)

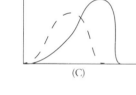 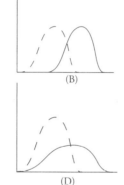

(A)

(B)

(C)

(D)

14. Nitrogen's electronegativity value is between those of phosphorus and oxygen. Which of the following correctly describes the relationship between the three values?

(A) The value for nitrogen is less than that of phosphorus because nitrogen is larger, but greater than that of oxygen because nitrogen has a greater effective nuclear charge.

(B) The value for nitrogen is less than that of phosphorus because nitrogen has less protons but greater than that of oxygen because nitrogen has less valence electrons.

(C) The value for nitrogen is greater than that of phosphorus because nitrogen has less electrons, but less than that of oxygen because nitrogen is smaller.

(D) The value for nitrogen is greater than that of phosphorus because nitrogen is smaller, but less than that of oxygen because nitrogen has a smaller effective nuclear charge.

15. A sample of a compound known to consist of only carbon, hydrogen, and oxygen is found to have a total mass of 29.05 g. If the mass of the carbon is 18.02 g and the mass of the hydrogen is 3.03 g, what is the empirical formula of the compound?

(A) C_2H_4O
(B) C_3H_6O
(C) $C_2H_6O_3$
(D) $C_3H_8O_2$

Questions 16-18 refer to the following.

A solution of carbonic acid, H_2CO_3, is titrated with sodium hydroxide. The following graph is produced:

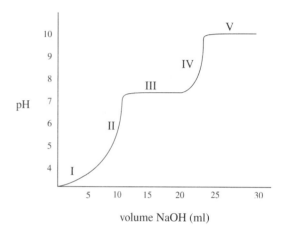

16. In addition to OH⁻, what species are present in the solution during section III of the graph?

(A) H_2CO_3, HCO_3^-, and CO_3^{2-}
(B) H_2CO_3 and HCO_3^-
(C) HCO_3^- and CO_3^{2-}
(D) H_2CO_3 and CO_3^2

17. What is the magnitude of the first dissociation constant?

(A) 10^{-2}
(B) 10^{-4}
(C) 10^{-6}
(D) 10^{-8}

GO ON TO THE NEXT PAGE.

18. If the concentration of the sodium hydroxide is increased prior to repeating the titration, what effect, if any, would that have on the graph?

    (A) The graph would not change at all.
    (B) The pH values at the equivalence points would increase.
    (C) The equivalence points would be reached with less volume of NaOH added.
    (D) The slope of the equivalence points would decrease.

19. Two solutions are mixed together, and the particulate representation below shows what is present after the reaction has gone to completion:

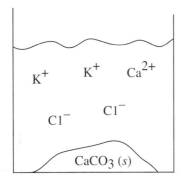

    Which of the two original solutions is the limiting reagent and why?

    (A) The potassium carbonate, because of the polyatomic anion.
    (B) The potassium carbonate, because there is no carbonate left after the reaction.
    (C) The calcium chloride, because there is an excess of calcium ions post-reaction.
    (D) The calcium chloride, because the component ions are smaller than those in potassium carbonate.

20. In which of the following circumstances is the value for K_{eq} always greater than 1?

	ΔH	ΔS
(A)	Positive	Positive
(B)	Positive	Negative
(C)	Negative	Negative
(D)	Negative	Positive

21. The structure of two oxoacids is shown below:

    $$H - \ddot{\underset{\cdot\cdot}{O}} - \ddot{\underset{\cdot\cdot}{Cl}}: \qquad H - \ddot{\underset{\cdot\cdot}{O}} - \ddot{\underset{\cdot\cdot}{F}}:$$

    Which would be a stronger acid, and why?

    (A) HOCl because the H–O bond is weaker than in HOF as chlorine is larger than fluorine.
    (B) HOCl because the H–O bond is stronger than in HOF as chlorine has a higher electronegativity than fluorine.
    (C) HOF because the H–O bond is stronger that in HOCl as fluorine has a higher electronegativity than chlorine.
    (D) HOF because the H–O bond is weaker than in HOCl as fluorine is smaller than chlorine.

22. During a chemical reaction, NO (g) gets reduced and no nitrogen-containing compound is oxidized. Which of the following is a possible product of this reaction?

    (A) NO_2 (g)
    (B) N_2 (g)
    (C) NO_3^- (aq)
    (D) NO_2^- (aq)

23. Which of the following pairs of substances would make a good buffer solution?

    (A) $HC_2H_3O_2$ (aq) and $NaC_2H_3O_2$ (aq)
    (B) H_2SO_4 (aq) and LiOH (aq)
    (C) HCl (aq) and KCl (aq)
    (D) HF (aq) and NH_3 (aq)

Questions 24-27 refer to the following.

Inside a calorimeter, 100.0 mL of 1.0 M hydrocyanic acid (HCN), a weak acid, and 100.0 mL of 0.50 M sodium hydroxide are mixed. The temperature of the mixture rises from 21.5 °C to 28.5 °C. The specific heat of the mixture is approximately 4.2 J/g°C, and the density is identical to that of water.

24. Identify the correct net ionic equation for the reaction that takes place.

    (A) HCN (aq) + OH$^-$ (aq) $\leftrightarrow$ CN$^-$ (aq) + H_2O (l)
    (B) HCN (aq) + NaOH (aq) $\leftrightarrow$ NaCN (aq) + H_2O (l)
    (C) H$^+$ (aq) + OH$^-$ (aq) $\leftrightarrow$ H_2O (l)
    (D) H$^+$ (aq) + CN$^-$ (aq) + Na$^+$ (aq) + OH$^-$ (aq) $\leftrightarrow$ H_2O (l) + CN$^-$ (aq) + Na$^+$ (aq)

GO ON TO THE NEXT PAGE.

25. What is the approximate enthalpy change for the reaction?

    (A) 1.5 kJ
    (B) 2.9 kJ
    (C) 5.9 kJ
    (D) 11.8 kJ

26. At ΔT increases, what happens to the equilibrium constant and why?

    (A) The equilibrium constant increases because more products are created.
    (B) The equilibrium constant increases because the rate of the forward reaction increases.
    (C) The equilibrium constant decreases because the equilibrium shifts to the left.
    (D) The value for the equilibrium constant is unaffected by temperature and will not change.

27. If the experiment is repeated with 200.0 mL of 1.0 M HCN and 100. mL of 0.50 M NaOH, what would happen to the values for ΔT and the ΔH?

	ΔT	ΔH_{rxn}
(A)	Increase	Increase
(B)	Stay the same	Stay the same
(C)	Decrease	Stay the same
(D)	Stay the same	Increase

28. The following diagrams show the Lewis structures of four different molecules. Which molecule would travel the farthest in a paper chromatography experiment using a polar solvent?

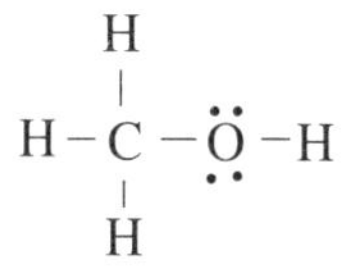

Methanol

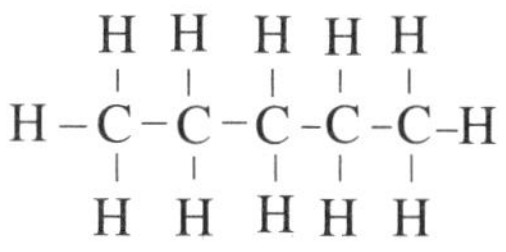

Pentane

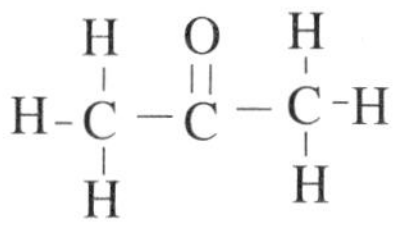

Acetone

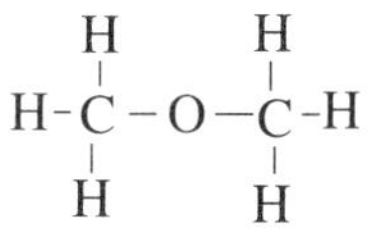

Ether

    (A) Methanol
    (B) Pentane
    (C) Acetone
    (D) Ether

29. The first ionization energy for a neutral atom of chlorine is 1.25 MJ/mol and the first ionization energy for a neutral atom of argon is 1.52 MJ/mol. How would the first ionization energy value for a neutral atom of potassum compare to those values?

    (A) It would be greater than both because potassium carries a greater nuclear charge then either chlorine or argon.
    (B) It would be greater than both because the size of a potassium atom is smaller than an atom of either chlorine or argon.
    (C) It would be less than both because there are more electrons in potassium, meaning they repel each other more effectively and less energy is needed to remove one.
    (D) It would be less than both because a valence electron of potassium is further from the nucleus than one of either chlorine or argon.

30. Which net ionic equation below represents a possible reaction that takes place when a strip of magnesium metal is oxidized by a solution of chromium (III) nitrate?

    (A) $Mg\ (s) + Cr(NO_3)_3\ (aq) \rightarrow Mg^{2+}(aq) + Cr^{3+}\ (aq) + 3\ NO_3^-\ (aq)$
    (B) $3\ Mg\ (s) + 2\ Cr^{3+} \rightarrow 3\ Mg^{2+} + 2\ Cr\ (s)$
    (C) $Mg\ (s) + Cr^{3+} \rightarrow Mg^{2+} + Cr\ (s)$
    (D) $3\ Mg\ (s) + 2\ Cr(NO_3)_3\ (aq) \rightarrow 3\ Mg^{2+}\ (aq) + 2\ Cr\ (s) +\ NO_3^-\ (aq)$

31. $PCl_3\ (g) + Cl_2\ (g) \leftrightarrow PCl_5\ (g)\quad \Delta H = -92.5$ kJ/mol

    In which of the following ways could the reaction above be manipulated to create more product?

    (A) Decreasing the concentration of PCl_3
    (B) Increasing the pressure
    (C) Increasing the temperature
    (D) None of the above

GO ON TO THE NEXT PAGE.

32. A pure solid substance is heated strongly. It first melts into a liquid, then boils and becomes a gas. Which of the following heating curves correctly shows the relationship between temperature and heat added?

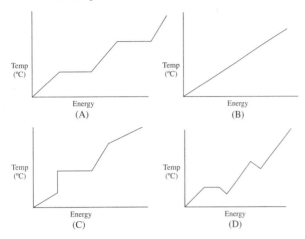

33. Consider the following reaction showing photosynthesis:

$$6\ CO_2\ (g) + 6\ H_2O\ (l) \rightarrow C_6H_{12}O_6\ (s) + 6\ O_2\ (g)$$
$$H = +\ 2800\ kJ/mol$$

Which of the following is true regarding the thermal energy in this system?

(A) It is transferred from the surroundings to the reaction.
(B) It is transferred from the reaction to the surroundings.
(C) It is transferred from the reactants to the products.
(D) It is transaction from the products to the reactants.

34.
$$SO_2Cl_2 \rightarrow SO_2\ (g) + Cl_2\ (g)$$

At 600 K, SO_2Cl_2 will decompose to form sulfur dioxide and chlorine gas via the above equation. If the reaction is found to be first order overall, which of the following will cause an increase in the half life of SO_2Cl_2?

(A) Increasing the initial concentration of SO_2Cl_2.
(B) Increasing the temperature at which the reaction occurs.
(C) Decreasing the overall pressure in the container.
(D) None of these will increase the half life.

Questions 35-38 refer to the following reaction.

$$2\ SO_2\ (g) + O_2\ (g) \rightarrow 2\ SO_3\ (g)$$

4.0 mol of gaseous SO_2 and 6.0 mol of O_2 gas are allowed to react in a sealed container.

35. Which particulate drawing best represents the contents of the flask after the reaction goes to completion?

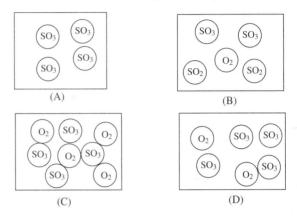

36. What percentage of the original pressure will the final pressure in the container be equal to?

(A)   67%
(B)   80%
(C)   100%
(D)   133%

37. At a given point in the reaction, all three gases are present at the same temperature. Which gas molecules will have the highest velocity and why?

(A) The O_2 molecules because they are the lightest.
(B) The O_2 molecules because they are the smallest.
(C) The SO_3 molecules because they are products in the reaction.
(D) Molecules of all three gases will have the same speed because they have the same temperature.

GO ON TO THE NEXT PAGE.

38. Under which of the following conditions would the gases in the container most deviate from ideal conditions and why?

    (A) Low pressures because the gas molecules would be spread far apart

    (B) High pressures because the gas molecules will be colliding frequently

    (C) Low temperatures because the intermolecular forces between the gas molecules would increase

    (D) High temperatures because the gas molecules are moving too fast to interact with each other

39. Four different acids are added to beakers of water, and the following diagrams represent the species present in each solution at equilibrium. Which acid has the highest pH?

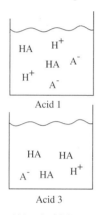

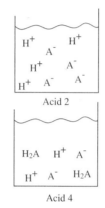

    (A) Acid 1
    (B) Acid 2
    (C) Acid 3
    (D) Acid 4

40. Which expression below should be used to calculate the mass of copper than can be plated out of a 1.0 M $Cu(NO_3)_2$ solution using a current of 0.75 A for 5.0 minutes?

    (A) $\dfrac{(5.0)(60)(0.75)(63.55)}{(96500)(2)}$

    (B) $\dfrac{(5.0)(60)(63.55)(2)}{(0.75)(96500)}$

    (C) $\dfrac{(5.0)(60)(96500)(0.75)}{(63.55)(2)}$

    (D) $\dfrac{(5.0)(60)(96500)(63.55)}{(0.75)(2)}$

41. Lewis diagrams for the nitrate and nitrite ions are shown above. Choose the statement that correctly describes the relationship between the two ions in terms of bond length and bond energy.

    (A) Nitrite has longer and stronger bonds than nitrate.
    (B) Nitrite has longer and weaker bonds than nitrate.
    (C) Nitrite has shorter and stronger bonds than nitrate.
    (D) Nitrite has shorter and weaker bonds than nitrate.

42. Examining data obtained from mass spectrometry supports which of the following?

    (A) the common oxidation states of elements
    (B) atomic size trends within the periodic table
    (C) ionization energy trends within the periodic table
    (D) the existence of isotopes

43. A 2.0 L flask holds 0.40 g of Helium gas. If the helium is evacuated into a larger container while the temperature is held constant, what will the effect on the entropy of the helium be?

    (A) It will remain constant as the number of helium molecules does not change.
    (B) It will decrease as the gas will be more ordered in the larger flask.
    (C) It will decrease because the molecules will collide with the sides of the larger flask less often than they did in the smaller flask.
    (D) It will increase as the gas molecules will be more dispersed in the larger flask.

GO ON TO THE NEXT PAGE.

Questions 44-48 refer to the following.

NO$_2$ gas is placed in a sealed, evacuated container and allowed to decompose via the following equation:

$$2 \, NO_2 \, (g) \leftrightarrow 2 \, NO \, (g) + O_2 \, (g)$$

The graph below indicates the change in concentration for each species over time.

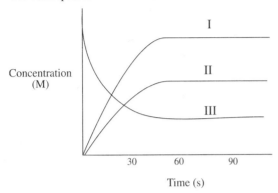

44. Using the numbers of the lines on the graph, identify which line belongs to which species.

	Line I	Line II	Line III
(A)	NO$_2$	NO	O$_2$
(B)	O$_2$	NO$_2$	NO
(C)	NO	O$_2$	NO$_2$
(D)	O$_2$	NO	NO$_2$

45. What is happening to the rate of the forward reaction at $t = 60$ s?

(A) It is increasing.
(B) It is decreasing.
(C) It is remaining constant.
(D) It has stopped.

46. At the reaction progresses, what happens to the value of the equilibrium constant K_p if the temperature remains constant?

(A) It stays constant.
(B) It increases exponentially.
(C) It increases linearly.
(D) It decreases exponentially.

47. What would happen to the slope of the NO$_2$ line if additional O$_2$ were injected into the container?

(A) It would increase, then level off.
(B) It would decrease, then level off.
(C) It would remain constant.
(D) It would increase, then decrease.

48. Using the graph, how could you determine the instantaneous rate of disappearance of NO$_2$ at $t = 30$ s?

(A) By determining the area under the graph at $t = 30$ s.
(B) By taking the slope of a line tangent to the NO$_2$ curve at $t = 30$ s.
(C) By using the values at $t = 30$ s and plugging them into the K_p expression.
(D) By measuring the overall gas pressure in the container at $t = 30$ s.

49. A proposed mechanism for a reaction is as follows:

$$NO_2 + F_2 \rightarrow NO_2F + F \qquad \text{Slow step}$$
$$F + NO_2 \rightarrow NO_2F \qquad \text{Fast step}$$

What is the order of the overall reaction?

(A) Zero order
(B) First order
(C) Second order
(D) Third order

50. Starting with a stock solution of 18.0 M H$_2$SO$_4$, what is the proper procedure to create a 1.00 L of a 3.0 M solution of H$_2$SO$_4$ in a volumetric flask?

(A) Add 167 mL of the stock solution to the flask, then fill the flask the rest of the way with distilled water while swirling the solution.
(B) Add 600 mL of the stock solution to the flask, then fill the flask the rest of the way with distilled water while swirling the solution.
(C) Fill the flask partway with water, then add 167 mL of the stock solution, swirling to mix it. Last, fill the flask the rest of the way with distilled water.
(D) Fill the flask partway with water, then add 600 mL of the stock solution, swirling to mix it. Last, fill the flask the rest of the way with distilled water.

GO ON TO THE NEXT PAGE.

51. The reaction shown in the above diagram is accompanied by a large increase in temperature. If all molecules shown are in their gaseous state, which statement accurately describes the reaction?

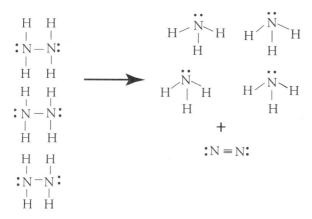

(A) It is an exothermic reaction in which entropy increases.
(B) It is an exothermic reaction in which entropy decreases.
(C) It is an endothermic reaction in which entropy increases.
(D) It is an endothermic reaction in which entropy decreases.

52. A student mixes equimolar amounts of KOH and $Cu(NO_3)_2$ in a beaker. Which of the following particulate diagrams correctly shows all species present after the reaction occurs?

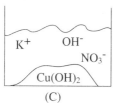

(A)

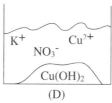

(B)

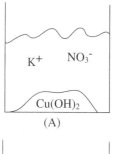

(C)

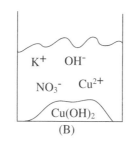

(D)

Questions 53-55 refer to the following.

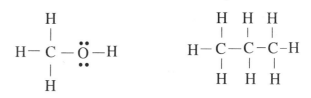

Methanol

Propane

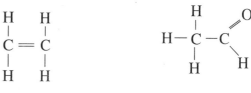

Ethene

Ethanal

53. Based on the strength of the intermolecular forces in each substance, estimate from greatest to smallest the vapor pressures of each substance in liquid state at the same temperature.

(A) propane > ethanal > ethene > methanol
(B) ethene > propane > ethanal > methanol
(C) ethanal > methanol > ethene > propane
(D) methanol > ethanal > propane > ethene

54. When in liquid state, which two substances are most likely to be miscible with water?

(A) propane and ethene
(B) methanol and propane
(C) ethene and ethanal
(D) methanol and ethanal

55. Between propane and ethene, which will likely have the higher boiling point and why?

(A) propane, because it has a greater molar mass
(B) propane, because it has a more polarizable electron cloud
(C) ethene, because of the double bond
(D) ethene, because it is smaller in size

GO ON TO THE NEXT PAGE.

56. $4 NH_3 (g) + 5 O_2 (g) \rightarrow 4 NO (g) + 6 H_2O (g)$

The above reaction will experience a rate increase by the addition of a cataylst such as platinum. Which of the following best explains why?

(A) The catalyst increases the overall frequency of collisions in the reactant molecules.
(B) The catalyst increases the frequency of collisions that occur at the proper orientation in the reactant molecules.
(C) The catalyst introduces a new reaction mechanism for the reaction.
(D) The catalyst increases the activation energy for the reaction.

57. The graph below shows the amount of potential energy between two hydrogen atoms as the distance between them changes. At which point in the graph would a molecule of H_2 be the most stable?

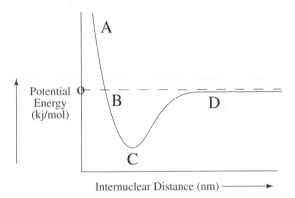

(A) Point A
(B) Point B
(C) Point C
(D) Point D

58. $N_2 (g) + O_2 (g) + Cl_2 (g) \leftrightarrow 2NOCl (g) \quad \Delta G = 132.6 \text{ kJ/mol}$

For the equilibrium above, what would happen to the value of ΔG if the concentration of N_2 were to increase and why?

(A) It would increase as the reaction would become more thermodynamically favored.
(B) It would increase as the reaction would shift right and create more products.
(C) It would decrease because there are more reactants present.
(D) It would stay the same because the value of K_{eq} would not change.

59. $C (s) + 2 S (s) \rightarrow CS_2 (l) \quad \Delta H = +92.0 \text{ kJ/mol}$

Which of the following energy level diagrams gives an accurate representation of the above reaction?

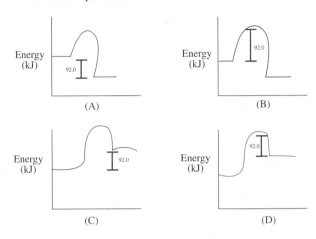

60. Two alloys are shown in the diagrams above—bronze, and steel. Which of the following correctly describes the malleability of both alloys compared to their primary metals?

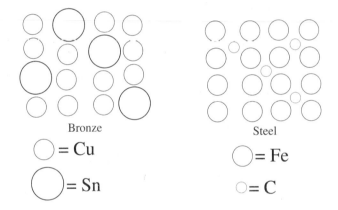

(A) Bronze's malleability would be comparable to that of copper, but steel's malleability would be significantly lower than that of iron.
(B) Bronze's malleability would be significantly higher than that of copper, but steel's malleability would be comparable to that of iron.
(C) Both bronze and steel would have malleability values similar to those of their primary metals.
(D) Both bronze and steel would have malleability values greater than those of their primary metals.

END OF SECTION I

CHEMISTRY
SECTION II
Time—1 hour and 30 minutes

<u>General Instructions</u>

Calculators, including those with programming and graphing capabilities, may be used in Part A. However, calculators with typewriter-style (QWERTY) keyboards are NOT permitted.

Pages containing a periodic table, the electrochemical series, and equations commonly used in chemistry will be available for your use.

You may write your answers with either a pen or a pencil. Be sure to write CLEARLY and LEGIBLY. If you make an error, you may save time by crossing it out rather than trying to erase it.

Write all your answers in the essay booklet. Number your answers as the questions are numbered in the examination booklet.

GO ON TO THE NEXT PAGE.

MATERIAL IN THE FOLLOWING TABLE AND IN THE TABLES ON THE NEXT 3 PAGES MAY BE USEFUL IN ANSWERING THE QUESTIONS IN THIS SECTION OF THE EXAMINATION.

PERIODIC CHART OF THE ELEMENTS

1 H 1.0																		2 He 4.0
3 Li 6.9	4 Be 9.0											5 B 10.8	6 C 12.0	7 N 14.0	8 O 16.0	9 F 19.0	10 Ne 20.2
11 Na 23.0	12 Mg 24.3											13 Al 27.0	14 Si 28.1	15 P 31.0	16 S 32.1	17 Cl 35.5	18 Ar 39.9
19 K 39.1	20 Ca 40.1	21 Sc 45.0	22 Ti 47.9	23 V 50.9	24 Cr 52.0	25 Mn 54.9	26 Fe 55.8	27 Co 58.9	28 Ni 58.7	29 Cu 63.5	30 Zn 65.4	31 Ga 69.7	32 Ge 72.6	33 As 74.9	34 Se 79.0	35 Br 79.9	36 Kr 83.8
37 Rb 85.5	38 Sr 87.6	39 Y 88.9	40 Zr 91.2	41 Nb 92.9	42 Mo 95.9	43 Tc (98)	44 Ru 101.1	45 Rh 102.9	46 Pd 106.4	47 Ag 107.9	48 Cd 112.4	49 In 114.8	50 Sn 118.7	51 Sb 121.8	52 Te 127.6	53 I 126.9	54 Xe 131.3
55 Cs 132.9	56 Ba 137.3	57 *La 138.9	72 Hf 178.5	73 Ta 180.9	74 W 183.9	75 Re 186.2	76 Os 190.2	77 Ir 192.2	78 Pt 195.1	79 Au 197.0	80 Hg 200.6	81 Tl 204.4	82 Pb 207.2	83 Bi 209.0	84 Po (209)	85 At (210)	86 Rn (222)
87 Fr (223)	88 Ra 226.0	89 †Ac 227.0															

*Lanthanum Series

58 Ce 140.1	59 Pr 140.9	60 Nd 144.2	61 Pm (145)	62 Sm 150.4	63 Eu 152.0	64 Gd 157.3	65 Tb 158.9	66 Dy 162.5	67 Ho 164.9	68 Er 167.3	69 Tm 168.9	70 Yb 173.0	71 Lu 175.0

†Actinium Series

90 Th 232.0	91 Pa 231.0	92 U 238.0	93 Np 237.0	94 Pu (244)	95 Am (243)	96 Cm (247)	97 Bk (247)	98 Cf (251)	99 Es (252)	100 Fm (258)	101 Md (258)	102 No (259)	103 Lr (260)

DO NOT DETACH FROM BOOK.

GO ON TO THE NEXT PAGE.

ADVANCED PLACEMENT CHEMISTRY EQUATIONS AND CONSTANTS

ATOMIC STRUCTURE

$$E = h\nu \qquad\qquad c = \lambda\nu$$

$$\lambda = \frac{h}{mv} \qquad\qquad p = mv$$

$$E_n = \frac{-2.178 \times 10^{-18}}{n^2} \text{ joule}$$

EQUILIBRIUM

$$K_a = \frac{[H^-][A^-]}{[HA]}$$

$$K_b = \frac{[OH^-][HB^+]}{[B]}$$

$$K_w = [OH^-][H^+] = 10^{-14} \text{ @ } 25^\circ C$$

$$= K_a \times K_b$$

$$pH = -\log[H^+], pOH = -\log[OH^-]$$

$$14 = pH + pOH$$

$$pH = pK_a + \log\frac{[A^-]}{[HA]}$$

$$pOH = pK_b + \log\frac{[HB^+]}{[B]}$$

$$pK_a = -\log K_a, pK_b = -\log K_b$$

$$K_p = K_c(RT)^{\Delta n},$$

where Δn = moles product gas – moles reactant gas

THERMOCHEMISTRY / KINETICS

$$\Delta S^\circ = \sum S^\circ_{\text{products}} - \sum S^\circ_{\text{reactants}}$$

$$\Delta H^\circ = \sum \Delta H^\circ_{f \text{ products}} - \sum \Delta H^\circ_{f \text{ reactants}}$$

$$\Delta G^\circ = \sum \Delta G^\circ_{f \text{ products}} - \sum \Delta G^\circ_{f \text{ reactants}}$$

$$\Delta G^\circ = \Delta H^\circ - T\Delta S^\circ$$

$$= -RT \ln K = -2.303 \, RT \log K$$

$$= -nFE^\circ$$

$$\Delta G = \Delta G^\circ + RT \ln Q = \Delta G^\circ + 2.303 \, RT \log Q$$

$$q = mc\Delta T$$

$$C_p = \frac{\Delta H}{\Delta T}$$

$$\ln[A]_t = -kt + \ln[A]_o$$

$$\frac{1}{[A]_t} = kt + \frac{1}{[A]_o}$$

$$\ln k = \frac{-E_a}{R}\left(\frac{1}{T}\right) + \ln A$$

E = energy v = velocity

ν = frequency n = principal quantum number

λ = wavelength m = mass

p = momentum

Speed of light, $c = 3.00 \times 10^8 \text{ m s}^{-1}$

Planck's constant, $h = 6.63 \times 10^{-34} \text{ J s}$

Boltzmann's constant, $k = 1.38 \times 10^{-23} \text{ J K}^{-1}$

Avogadro's number $= 6.022 \times 10^{23} \text{ mol}^{-1}$

Electron charge, $e = -1.602 \times 10^{-19} \text{ coulomb}$

1 electron volt/atom $= 96.5 \text{ kJ mol}^{-1}$

Equilibrium Constants

K_a (weak acid)
K_b (weak base)
K_w (water)
K_p (gas pressure)
K_c (molar concentrations)

S° = standard entropy
H° = standard enthalpy
G° = standard free energy
E° = standard reduction potential
T = temperature
n = moles
m = mass
q = heat
c = specific heat capacity
C_p = molar heat capacity at constant pressure
E_a = activation energy
k = rate constant
A = frequency factor

Faraday's Constant, F = 96,500 coulombs per mole
of electrons

Gas Constant, R = 8.31 J mol^{-1} K^{-1}
= 0.0821 L atm mol^{-1} K^{-1}
= 8.31 volt coulomb mol^{-1} K^{-1}

GO ON TO THE NEXT PAGE.

ADVANCED PLACEMENT CHEMISTRY EQUATIONS AND CONSTANTS

GASES, LIQUIDS, AND SOLUTIONS

$$PV = nRT$$

$$\left(P + \frac{n^2 a}{V^2}\right)(V - nb) = nRT$$

$$P_A = P_{total} \cdot X_A, \text{ where } X_A = \frac{\text{moles A}}{\text{total moles}}$$

$$P_{total} = P_A + P_B + P_C{}^+ \ldots$$

$$n = \frac{m}{M}$$

$$K = {}^\circ C + 273$$

$$\frac{P_1 V_1}{T_1} = \frac{P_2 V_2}{T_2}$$

$$D = \frac{m}{V}$$

$$U_{rms} = \sqrt{\frac{3kT}{m}} = \sqrt{\frac{3RT}{M}}$$

$$KE \text{ per molecule} = \frac{1}{2} mv^2$$

$$KE \text{ per mole} = \frac{3}{2} RT$$

$$\frac{r_1}{r_2} = \sqrt{\frac{M_2}{M_1}}$$

$$\text{molarity, } M = \text{ moles solute per liter solution}$$
$$\text{molality} = \text{ moles solute per kilogram solvent}$$
$$\Delta T_f = iK_f \times \text{molality}$$
$$\Delta T_b = iK_b \times \text{molality}$$
$$\pi = MRT$$
$$A = abc$$

P = pressure
V = volume
T = temperature
n = number of moles
D = density
m = mass
v = velocity

U_{rms} = root-mean-square speed
KE = kinetic energy
r = rate of effusion
M = molar mass
π = osmotic pressure
i = van't Hoff factor
K_f = molal freezing-point depression constant
K_b = molal boiling-point elevation constant
A = absorbance
a = molar absorptivity
b = path length
c = concentration
Q = reaction quotient
l = current (amperes)
q = charge (coulombs)
t = time (seconds)
E° = standard reduction potential
K = equilibrium constant

Gas constant, R = 8.31 J mol^{-1}K^{-1}
= 0.0821 L atm mol^{-1} K^{-1}
= 8.31 volt coulomb mol^{-1} K^{-1}
Boltzmann's constant, k = 1.38×10^{-23} J K^{-1}
K_f for H_2O = 1.86 K kg mol^{-1}
K_b for H_2O = 0.512 K kg mol^{-1}
1 atm = 760 mm Hg
= 760 torr
STP = 0.000°C and 1.000 atm
1 faraday, F = 96,500 coulumbs per mole
of electrons

OXIDATION REDUCTION; ELECTROCHEMISTRY

$$Q = \frac{[C]^c [D]^d}{[A]^a [B]^b} \text{ where } a\,A + b\,B \rightarrow c\,C + d\,D$$

$$I = \frac{q}{t}$$

$$E_{cell} = E^\circ{}_{cell} - \frac{RT}{nF} \ln Q = E^\circ{}_{cell} - \frac{0.0592}{n} \log Q \text{ @ 25°C}$$

$$\log K = \frac{nE^\circ}{0.0592}$$

GO ON TO THE NEXT PAGE.

STANDARD REDUCTION POTENTIALS IN AQUEOUS SOLUTION AT 25 C (in V)			
$F_2(g) + 2\,e^-$	$\rightarrow$	$2\,F^-$	2.87
$Co^{3+} + e^-$	$\rightarrow$	Co^{2+}	1.82
$Au^{3+} + 3e^-$	$\rightarrow$	$Au(s)$	1.50
$Cl_2(g) + 2\,e^-$	$\rightarrow$	$2\,Cl^-$	1.36
$O_2(g) + 4H^+ + 4\,e^-$	$\rightarrow$	$2\,H_2O$	1.23
$Br_2(l) + 2\,e^-$	$\rightarrow$	$2\,Br^-$	1.07
$2\,Hg^{2+} + 2\,e^-$	$\rightarrow$	Hg_2^{2+}	0.92
$Hg^{2+} + 2\,e^-$	$\rightarrow$	$Hg(l)$	0.85
$Ag^+ + e^-$	$\rightarrow$	$Ag(s)$	0.80
$Hg_2^{2+} + 2\,e^-$	$\rightarrow$	$2\,Hg(l)$	0.79
$Fe^{3+} + e^-$	$\rightarrow$	Fe^{2+}	0.77
$I_2(s) + 2\,e^-$	$\rightarrow$	$2\,I^-$	0.53
$Cu^+ + e^-$	$\rightarrow$	$Cu(s)$	0.52
$Cu^{2+} + 2\,e^-$	$\rightarrow$	$Cu(s)$	0.34
$Cu^{2+} + e^-$	$\rightarrow$	Cu^+	0.15
$Sn^{4+} + 2\,e^-$	$\rightarrow$	Sn^{2+}	0.15
$S(s) + 2\,H^+ + 2\,e^-$	$\rightarrow$	H_2S	0.14
$2\,H^+ + 2\,e^-$	$\rightarrow$	$H_2(g)$	0.00
$Pb^{2+} + 2\,e^-$	$\rightarrow$	$Pb(s)$	−0.13
$Sn^{2+} + 2\,e^-$	$\rightarrow$	$Sn(s)$	−0.14
$Ni^{2+} + 2\,e^-$	$\rightarrow$	$Ni(s)$	−0.25
$Co^{2+} + 2\,e^-$	$\rightarrow$	$Co(s)$	−0.28
$Tl^+ + e^-$	$\rightarrow$	$Tl(s)$	−0.34
$Cd^{2+} + 2\,e^-$	$\rightarrow$	$Cd(s)$	−0.40
$Cr^{3+} + e^-$	$\rightarrow$	Cr^{2+}	−0.41
$Fe^2 + 2\,e^-$	$\rightarrow$	$Fe(s)$	−0.44
$Cr^{3+} + 3\,e^-$	$\rightarrow$	$Cr(s)$	−0.74
$Zn^{2+} + 2\,e^-$	$\rightarrow$	$Zn(s)$	−0.76
$Mn^{2+} + 2\,e^-$	$\rightarrow$	$Mn(s)$	−1.18
$Al^{3+} + 3\,e^-$	$\rightarrow$	$Al(s)$	−1.66
$Be^{2+} + 2\,e^-$	$\rightarrow$	$Be(s)$	−1.70
$Mg^{2+} + 2\,e^-$	$\rightarrow$	$Mg(s)$	−2.37
$Na^+ + e^-$	$\rightarrow$	$Na(s)$	−2.71
$Ca^{2+} + 2\,e^-$	$\rightarrow$	$Ca(s)$	−2.87
$Sr^{2+} + 2\,e^-$	$\rightarrow$	$Sr(s)$	−2.89
$Ba^{2+} + 2\,e^-$	$\rightarrow$	$Ba(s)$	−2.90
$Rb^+ + e^-$	$\rightarrow$	$Rb(s)$	−2.92
$K^+ + e^-$	$\rightarrow$	$K(s)$	−2.92
$Cs^+ + e^-$	$\rightarrow$	$Cs(s)$	−2.92
$Li^+ + e^-$	$\rightarrow$	$Li(s)$	−3.05

GO ON TO THE NEXT PAGE.

CHEMISTRY

Section II

(Total time—90 minutes)

Part A

YOU MAY USE YOUR CALCULATOR FOR PART A

THE METHODS USED AND THE STEPS INVOLVED IN ARRIVING AT YOUR ANSWERS MUST BE SHOWN CLEARLY. It is to your advantage to do this since you may obtain partial credit if you do, and you will receive little or no credit if you do not. Attention should be paid to significant figures.

1. A student is tasked with determining the identity of an unknown carbonate compound with a mass of 1.89 g. The compound is first placed in water, where it dissolves completely. The K_{sp} value for several carbonate-containing compounds are given below.

Compound	K_{sp}
Lithium Carbonate	8.15×10^{-4}
Nickel (II) Carbonate	1.42×10^{-7}
Strontium Carbonate	5.60×10^{-10}

    (a)   In order to precipitate the maximum amount of the carbonate ions from solution, which of the following should be added to the carbonate solution: $LiNO_3$, $Ni(NO_3)_2$, or $Sr(NO_3)_2$? Justify your answer.

    (b)   For the carbonate compound that contains the cation chosen in part (a), determine the concentration of each ion of that compound in solution at equilibrium.

    (c)   When mixing the solution, should the student ensure the carbonate solution or the nitrate solution is in excess? Justify your answer.

    (d)   After titrating sufficient solution to precipitate out all of the carbonate ions, the student filters the solution before placing it in a crucible and heating it to drive off the water. After several heatings, he final mass of the precipitate remains constant and is determined to be 2.02 g.

        (i)   Determine the number of moles of precipitate.

        (ii) Determine the mass of carbonate present in the precipitate.

    (e)   Determine the percent, by mass, of carbonate in the original sample.

    (f)   Is the original compound most likely lithium carbonate, sodium carbonate, or potassium carbonate? Justify your answer.

GO ON TO THE NEXT PAGE.

2. The **unbalanced** reaction between potassium permanganate and acidified iron (II) sulfate is a redox reaction that proceeds as follows:

$$H^+ (aq) + Fe^{2+} (aq) + MnO_4^- (aq) \rightarrow Mn^{2+} (aq) + Fe^{3+} (aq) + H_2O (l)$$

(a) Provide the equations for both half reactions that occur below:
   (i) Oxidation half reaction.
   (ii) Reduction half reaction.
(b) What is the balanced net ionic equation?

A solution of 0.150 M potassium permanganate, a dark purple solution, is placed in a buret before being titrated into a flask containing 50.00 mL of iron (II) sulfate solution of unknown concentration. The following data describes the colors of the various ions in solution:

Ion	Color in solution
H^+	Colorless
Fe^{2+}	Pale Green
MnO_4^-	Dark Purple
Mn^{2+}	Colorless

Ion	Color in Solution
Fe^{3+}	Yellow
K^+	Colorless
SO_4^{2-}	Colorless

(c) Describe the color of the solution in the flask at the following points:
   (i) Before titration begins
   (ii) During titration prior to the endpoint
   (iii) At the endpoint of the titration
(d) (i) If 15.55 mL of permanganate are added to reach the endpoint, what is the initial concentration of the iron (II) sulfate?
   (ii) The actual concentration of the $FeSO_4$ is 0.250 M. Calculate the percent error.
(e) Could the following errors have led to the experimental result deviating in the direction that it did? You must justify your answers quantitatively.
   (i) 55.0 mL of $FeSO_4$ was added to the flask prior to titration instead of 50.0 mL.
   (ii) The concentration of the potassium permanganate was actually 0.160 M instead of 0.150 M.

GO ON TO THE NEXT PAGE.

3. Propane gas can be burned in air via the following reaction:

$$2 \, NH_3 \, (g) + 3 \, N_2O \, (g) \rightarrow 4 \, N_2 \, (g) + 3 \, H_2O \, (g)$$

The standard entropy of formation values for the varying substances are listed in the table below.

Substance	$S°$ (J/molK)
$NH_3 \, (g)$	193
$N_2O \, (g)$	220
$N_2 \, (g)$	192
$H_2O \, (g)$	189

(a) Calculate the entropy value for the overall reaction

Several bond enthalpies are listed in the table below.

Bond	Enthalpy (kJ/mol)	Bond	Enthalpy (kJ/mol)
N–H	388	N=N	409
N–O	210	N≡N	941
N=O	630	O–H	463

(b) Calculate the enthalpy value for the overall reaction
(c) Is this reaction thermodynamically favored at 25°C? Justify your answer.
(d) If 25.00 g of NH_3 reacts with 25.00 g of N_2O:
   (i) Will energy be released or absorbed?
   (ii) What is the magnitude of the energy change?
(e) On the reaction coordinates below, draw a line showing the progression of this reaction. Label both ΔH and E_a on the graph.

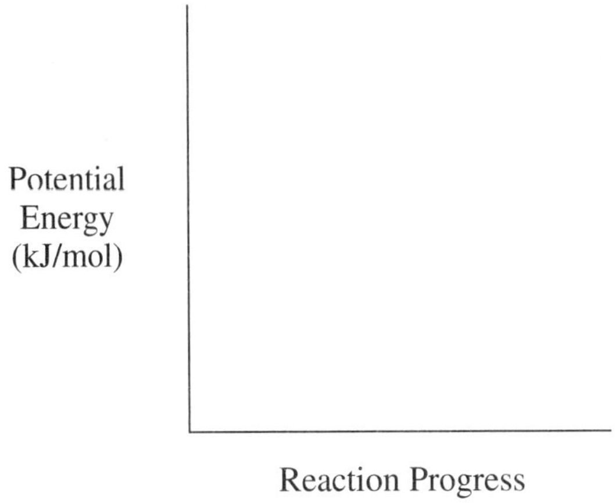

Potential
Energy
(kJ/mol)

Reaction Progress

GO ON TO THE NEXT PAGE.

Part B

CALCULATORS MAY BE USED FOR PART B

4. The acetyl ion has a formula of $C_2H_3O^-$ and two possible Lewis electron-dot diagram representations:

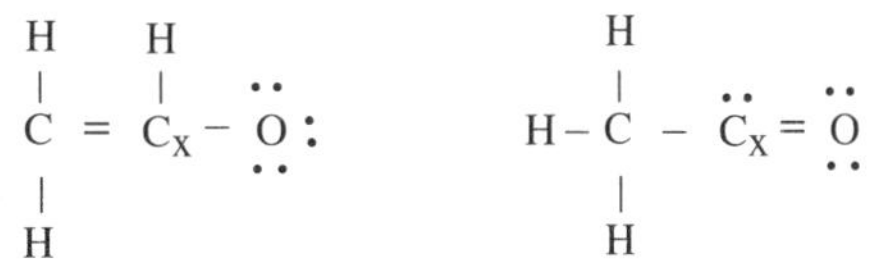

   (a)   Using formal charge, determine which structure is the most likely correct structure and circle it.

   (b)   For carbon atom "x" in the structure you chose:

        (i)   What is the hybridization around the atom?

        (ii) How many sigma and pi bonds has the atom formed?

   (c)   A hydrogen ion attaches itself to the the acetyl ion, creating C_2H_4O. Draw the Lewis diagram of the new molecule.

5. A student tests the conductivity of three different acid samples, each with a concentration of 0.5 M and a volume of 100.0 mL. The conductivity was recorded in microsiemens per centimeter in the table below:

Sample	Conductivity (μS/cm)
1	26820
2	8655
3	35120

   (a)   The three acids are known to be HCl, H_2SO_4, and H_3PO_4. Identify which sample is which acid. Justify your answer.

   (b)   The HCl solution is then titrated with a 1.50 M solution of the weak base methylamine, CH_3NH_2. ($K_b = 4.38 \times 10^{-4}$)

        (i)   Write out the net ionic equation for this reaction.

        (ii) Determine the pH of the solution after 20.0 mL of methylamine has been added.

GO ON TO THE NEXT PAGE.

6. The photoelectron spectrum of an element is given below:

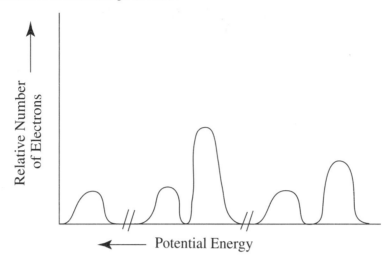

(a) Using your knowledge of atomic structure, explain the following:
   (i) The reason for the three discrete areas of ionization energies
   (ii) The justification for there being a total of five peaks
   (iii) The relative heights of the peaks when compared to each other
(b) Identify the element this spectra most likely belongs to and write out its full electron configuration.

7. Hydrazine (N_2H_4) can be produced commercially via the Raschig process. The following is a proposed mechanism:

Step 1: $NH_3 (aq) + OCl^- (aq) \rightarrow NH_2Cl (aq) + OH^-$
Step 2: $NH_2Cl (aq) + NH_3 (aq) \rightarrow N_2H_5^+ (aq) + Cl^-$
Step 3: $N_2H_5^+ (aq) + OH^- \rightarrow N_2H_4 (aq) + H_2O (l)$

(a) (i) What is the equation for the overall reaction?
     (ii) Identify any catalysts or intermediates from the reaction mechanism.
(b) The rate law for the reaction is determined to be rate = $k[NH_3][OCl^-]$
     (i) Which elementary step is the slowest one? Justify your answer.
     (ii) If the reaction is measured over the course of several minutes, what would the units on the rate constant be?

STOP

END OF EXAM

_____

Chapter 11
Practice Test 1
Answers and
Explanations

ANSWER KEY EXAM 1

1.	A	31.	B
2.	D	32.	A
3.	D	33.	A
4.	C	34.	D
5.	D	35.	C
6.	C	36.	B
7.	B	37.	A
8.	B	38.	C
9.	C	39.	C
10.	A	40.	A
11.	D	41.	C
12.	A	42.	D
13.	D	43.	D
14.	D	44.	C
15.	B	45.	C
16.	C	46.	A
17.	B	47.	A
18.	C	48.	B
19.	B	49.	C
20.	D	50.	C
21.	D	51.	A
22.	B	52.	D
23.	A	53.	B
24.	A	54.	D
25.	C	55.	B
26.	C	56.	C
27.	C	57.	C
28.	A	58.	D
29.	D	59.	C
30.	B	60.	A

QUESTIONS	EXPLANATIONS

SECTION I—MULTIPLE CHOICE

1. $2\ ClF\ (g) + O_2\ (g) \leftrightarrow Cl_2O\ (g) + F_2O\ (g)\ \ \Delta H = 167\ kJ$

During the reaction above, the product yield can be increased by increasing the temperature of the reaction. Why is this effective?

(A) The reaction is endothermic, therefore adding heat will shift it to the right.

(B) Increasing the temperature increases the speed of the molecules, meaning there will be more collisions between them.

(C) The reactants are less massive than the products, and an increase in temperature will cause their kinetic energy to increase more than that of the products.

(D) The increase in temperature allows for a higher percentage of molecular collisions to occur with the proper orientation to create the product.

1. **A** The reaction is endothermic, meaning energy is a reactant (appears on the left side of the equation). Adding stress to the left side increases the rate of the forward reaction, creating more products.

2. $CH_3NH_2\ (aq) + H_2O\ (l) \leftrightarrow OH^-\ (aq) + CH_3NH_3^+\ (aq)$

The above reaction represents the reaction between the base methylamine ($K_b = 4.38 \times 10^{-4}$) and water. Which of the following best represents the concentrations of the various species at equilibrium?

(A) $[OH^-] > [CH_3NH_2] = [CH_3NH_3^+]$

(B) $[OH^-] = [CH_3NH_2] = [CH_3NH_3^+]$

(C) $[CH_3NH_2] > [OH^-] > [CH_3NH_3^+]$

(D) $[CH_3NH_2] > [OH^-] = [CH_3NH_3^+]$

2. **D** Weak bases do not ionize fully in solution, and most of the methylamine molecules will not deprotonate. The hydroxide and conjugate acid ions are created in a 1:1 ratio and therefore will be equal.

3. The diagram below shows the relative atomic sizes of three different elements from the same period. What of the following must be true about the elements?

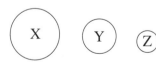

(A) The effective nuclear charge would be the greatest in element X.

(B) The first ionization energy will be greatest in element X.

(C) The electron shielding effect will be greatest in element Z.

(D) The electronegativity value will be greatest in element Z.

3. **D** Moving across a period, atomic size decreases. Therefore, atom Z will be furthest to the right (have the most protons), and thus have the highest electronegativity value.

QUESTIONS	EXPLANATIONS

4. A sealed, rigid container contains three gases: 28.0 g of nitrogen, 40.0 g of argon, and 36.0 g of water vapor. If the total pressure exerted by the gases is 2.0 atm, what is the partial pressure of the nitrogen?

   (A) 0.33 atm
   (B) 0.40 atm
   (C) 0.50 atm
   (D) 2.0 atm

4. **C** There are 1 mole of N_2, 1 mole of Ar, and 2 moles of water in the container. The mole fraction of nitrogen is: $1/4 = 0.25$. $P_{N2} = X_{N2})((P_{total})$ $P_{N2} = (0.25)(2.0) = 0.50$ atm

5. A sample of liquid NH_3 is brought to its boiling point. Which of the following occurs during the boiling process?

   (A) The N-H bonds within the NH_3 molecules together break apart.
   (B) The overall temperature of the solution rises as the NH_3 molecules speed up.
   (C) The amount of energy within the system remains constant.
   (D) The hydrogen bonds holding separate NH_3 molecules together break apart.

5. **D** When a covalent substance undergoes a phase change, the bonds between the various molecules inside the substance are what break apart.

Questions 6-10 refer to the following.

Two half-cells are set up as follows:

Half-Cell A: Strip of Cu (s) in $CuNO_3$ (aq)
Half-Cell B: Strip of Zn (s) in $Zn(NO_3)_2$ (aq)

When the cells are connected according to the diagram below, the following reaction occurs:

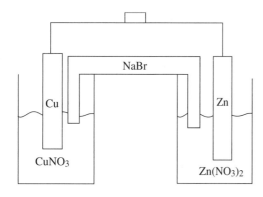

$2 Cu^+$ (aq) + Zn (s) → 2 Cu (s) + Zn^{2+} (aq) $E° = +1.28$ V

6. Correctly identify the anode and cathode in this reaction, as well as where oxidation and reduction are taking place.

   (A) Cu is the anode where oxidation occurs, and Zn is the cathode where reduction occurs.
   (B) Cu is the anode where reduction occurs, and Zn is the cathode where oxidation occurs.
   (C) Zn is the anode where oxidation occurs, and Cu is the cathode where reduction occurs.
   (D) Zn is the anode where reduction occurs, and Cu is the cathode where oxidation occurs.

6. **C** Copper's oxidation state changes from +1 to 0, meaning it has gained electrons and is being reduced, and reduction occurs at the cathode. Zinc's oxidation state changes from 0 to +2, meaning it has lost electrons and is being oxidized, which occurs at the anode.

7. How many moles of electron must be transferred to create 127 g of copper?

   (A) 1 mole of electrons
   (B) 2 moles of electrons
   (C) 3 moles of electrons
   (D) 4 moles of electrons

7. **B** 127 g is 2 moles of copper, which is what appears on the balanced equation. To change one mole of copper from +1 to 0, 1 mole of electrons is required. Twice as many moles being created means twice as many electrons are needed.

QUESTIONS	EXPLANATIONS

8. If the $Cu^+ + e^- \rightarrow Cu$ (s) half reaction has a standard reduction potential of +0.52 V, what is the standard reduction potential for the $Zn^{2+} + 2e^- \rightarrow Zn$ (s) half reaction?

   (A) +0.76 V
   (B) −0.76 V
   (C) +0.24 V
   (D) −0.24 V

8. **B** $E_{cell} = E_{red} + E_{ox}$

   $1.28\ \text{V} = 0.52\ \text{V} + E_{ox}$

   $E_{ox} = 0.76\ \text{V} \qquad -E_{ox} = E_{red}$

   $E_{red} = -0.76\ \text{V}$

9. As the reaction progresses, what will happen to the overall voltage of the cell?

   (A) It will increase as $[Zn^{2+}]$ increases.
   (B) It will increase as $[Cu^+]$ increases.
   (C) It will decrease as $[Zn^{2+}]$ increases.
   (D) The voltage will remain constant.

9. **C** As the reaction progresses, $[Cu^+]$ will decrease and $[Zn^{2+}]$ will increase. With a lower concentration, the number of electrons that are being transferred will decrease, decreasing the overall potential of the reaction.

10. What will happen in the salt bridge as the reaction progresses?

    (A) The Na^+ ions will flow to the Cu/Cu^+ half-cell.
    (B) The Br^- ions will flow to the Cu/Cu^+ half-cell.
    (C) Electrons will transfer from the Cu/Cu^+ half-cell to the Zn/Zn^{2+} half-cell.
    (D) Electrons will transfer from the Zn/Zn^{2+} half-cell to the Cu/Cu^+ half-cell.

10. **A** The electron transfer does not happen across the salt bridge, eliminating options (C) and (D). As the reaction progresses and $[Cu^+]$ decreases in the copper half-cell, positively charged sodium ions are transferred in to keep the charge balanced within the half-cell.

11. For a reaction involving nitrogen monoxide inside a sealed flask, the value for the reaction quotient (Q) was found to be 1.1×10^2 at a given point. If, after this point, the amount of NO gas in the flask increased, which reaction is most likely taking place in the flask?

    (A) $NOBr\ (g) \leftrightarrow NO\ (g) + ½\ Br_2\ (g) \qquad K_c = 3.4 \times 10^{-2}$
    (B) $2\ NOCl\ (g) \leftrightarrow 2\ NO\ (g) + Cl_2\ (g) \qquad K_c = 1.6 \times 10^{-5}$
    (C) $2\ NO\ (g) + 2\ H_2\ (g) \leftrightarrow N_2\ (g) + 2\ H_2O\ (g)$
    $K_c = 4.0 \times 10^6$
    (D) $N_2\ (g) + O_2\ (g) \leftrightarrow 2\ NO\ (g) \qquad K_c = 4.2 \times 10^2$

11. **D** When $Q > K_c$, the numerator of the equilibrium expression (the product concentration) is too big, and the equation shifts to the left. This is true for both (A) and (B), meaning [NO] would decrease. When $Q < K_c$, the numerator/product concentrations need to increase. This is the case in (C) and (D), but NO (g) is only a product in (D).

| **QUESTIONS** | **EXPLANATIONS** |

12. Which of the following substances has an asymmetrical molecular structure?

    (A) SF_4
    (B) PCl_5
    (C) BF_3
    (D) CO_2

12. **A**

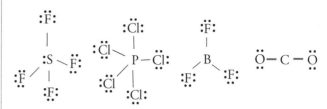

The only tricky bit here it to remember that boron is considered stable with only six electrons in its valence shell.

13.

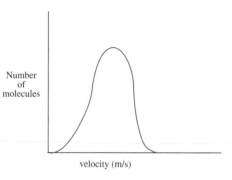

Number of molecules

velocity (m/s)

The diagram above shows the speed distribution of molecules in a gas held at 200 K. Which of the following representations would best represent the gas at a higher temperature? (Note: The original line is shown as a dashed line in the answer options.)

13. **D** At a higher temperature, the average velocity of the gas molecules would be greater. Additionally, they would have a greater spread of potential velocities, which would lead to a wider curve.

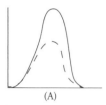

(A)

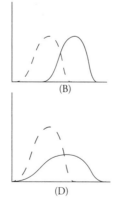

(B)

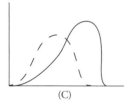

(C)

(D)

14. Nitrogen's electronegativity value is between those of phosphorus and oxygen. Which of the following correctly describes the relationship between the three values?

(A) The value for nitrogen is less than that of phosphorus because nitrogen is larger, but greater than that of oxygen because nitrogen has a greater effective nuclear charge.

(B) The value for nitrogen is less than that of phosphorus because nitrogen has less protons but greater than that of oxygen because nitrogen has less valence electrons.

(C) The value for nitrogen is greater than that of phosphorus because nitrogen has less electrons, but less than that of oxygen because nitrogen is smaller.

(D) The value for nitrogen is greater than that of phosphorus because nitrogen is smaller, but less than that of oxygen because nitrogen has a smaller effective nuclear charge.

14. **D** Nitrogen only has two shells of electrons while phosphorus has three, making nitrogen smaller and more able to attract additional electrons, meaning a higher electronegativity. Nitrogen and oxygen both have two shells, but oxygen has more protons and an effective nuclear charge of +6 vs. nitrogen's effective nuclear charge of +5. Thus, oxygen has a higher electronegativity.

15. A sample of a compound known to consist of only carbon, hydrogen, and oxygen is found to have a total mass of 29.05 g. If the mass of the carbon is 18.02 g and the mass of the hydrogen is 3.03 g, what is the empirical formula of the compound?

(A) C_2H_4O
(B) C_3H_6O
(C) $C_2H_6O_3$
(D) $C_3H_8O_2$

15. **B** First, the mass of the oxygen must be calculated: 29.05 g – 18.02 g – 3.03 g = 8.00 g.

Converting **each** of those to moles yield 0.5 moles of oxygen, 1.5 moles of carbon, and 3.0 moles of hydrogen. Thus, for every one oxygen atom there are three carbon atoms and six hydrogen atoms.

Questions 16-18 refer to the following.

A solution of carbonic acid, H_2CO_3, is titrated with sodium hydroxide. The following graph is produced:

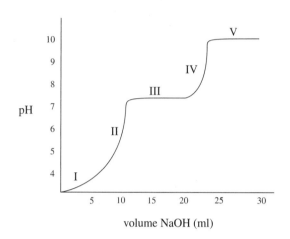

volume NaOH (ml)

16. In addition to OH^-, what species are present in the solution during section III of the graph?

    (A) H_2CO_3, HCO_3^-, and CO_3^{2-}
    (B) H_2CO_3 and HCO_3^-
    (C) HCO_3^- and CO_3^{2-}
    (D) H_2CO_3 and CO_3^2

16. **C** During sections I-II, the following reaction occurs: H_2CO_3 (aq) + OH^- (aq) ↔ HCO_3^- (aq) + H_2O (l). The endpoint of that is reached when all H_2CO_3 has reacted, meaning that in sections III and IV the following occurs: HCO_3^- (aq) + OH^- (aq) ↔ CO_3^{2-} (aq) + H_2O (l).

17. What is the magnitude of the first dissociation constant? (10^{-4})

    (A) 10^{-2}
    (B) 10^{-4}
    (C) 10^{-6}
    (D) 10^{-8}

17. **B** $pH = pK_a + [HCO_3^-]/[H_2CO_3]$. At a point in the graph where half of all the acid has reacted, the last part of the Henderson-Hasselbalch equation cancels out, leaving $pH = pK_a$. The first equivalence point occurs at a volume of 10 mL, and thus the half-equivalence point is at a volume of 5 mL. The pH at this point is 4, so: $4 = -\log K_a$. K_a is somewhere around 10^{-4}.

QUESTIONS	EXPLANATIONS

18. If the concentration of the sodium hydroxide is increased prior to repeating the titration, what effect, if any, would that have on the graph?

   (A) The graph would not change at all.
   (B) The pH values at the equivalence points would increase.
   (C) The equivalence points would be reached with less volume of NaOH added.
   (D) The slope of the equivalence points would decrease.

18. **C** A more concentrated NaOH solution means more moles of NaOH are added per drop, so a lower volume of NaOH would be needed to add enough moles to reach the equivalence point.

19. Two solutions are mixed together, and the particulate representation below shows what is present after the reaction has gone to completion:

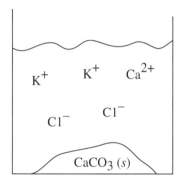

Which of the two original solutions is the limiting reagent and why?

   (A) The potassium carbonate, because of the polyatomic anion.
   (B) The potassium carbonate, because there is no carbonate left after the reaction.
   (C) The calcium chloride, because there is an excess of calcium ions post-reaction.
   (D) The calcium chloride, because the component ions are smaller than those in potassium carbonate.

19. **B** If extra Ca^{2+} ions are in solution, that means there were not enough CO_3^{2-} ions present for the Ca^{2+} ions to fully react.

QUESTIONS	EXPLANATIONS

20. In which of the following circumstances is the value for K_{eq} always greater than 1?

	ΔH	ΔS
(A)	Positive	Positive
(B)	Positive	Negative
(C)	Negative	Negative
(D)	Negative	Positive

20. **D** Via $K = e^{-\Delta G/RT}$, if ΔG is negative the value for K will be greater than one. Via $\Delta G = \Delta H - T\Delta S$, ΔG is always negative when ΔH is negative and ΔS is positive.

21. The structure of two oxoacids is shown below:

$$H - \ddot{\underset{\cdot\cdot}{O}} - \ddot{\underset{\cdot\cdot}{Cl}} : \qquad H - \ddot{\underset{\cdot\cdot}{O}} - \ddot{\underset{\cdot\cdot}{F}} :$$

Which would be a stronger acid, and why?

(A) HOCl because the H–O bond is weaker than in HOF as chlorine is larger than fluorine.
(B) HOCl because the H–O bond is stronger than in HOF as chlorine has a higher electronegativity than fluorine.
(C) HOF because the H–O bond is stronger that in HOCl as fluorine has a higher electronegativity than chlorine.
(D) HOF because the H–O bond is weaker than in HOCl as fluorine is smaller than chlorine.

21. **D** The weaker the O-H bond is in an oxoacid, the stronger the acid will be as the H^+ ions are more likely to dissociate. The O-F bond in HOF is stronger than the O-Cl bond in HOCl because fluorine is smaller (and thus more electronegative) than chlorine. If the O-F bond is stronger, the O-H bond is correspondingly weaker, making HOF the stronger acid.

22. During a chemical reaction, NO (g) gets reduced and no nitrogen-containing compound is oxidized. Which of the following is a possible product of this reaction?

(A) NO_2 (g)
(B) N_2 (g)
(C) NO_3^- (aq)
(D) NO_2^- (aq)

22. **B** The oxidation state of nitrogen in NO is +2. If the nitrogen is reduced, that value must get more negative. The oxidation state of nitrogen in N_2 is 0, so that fits the bill. The oxidation state on the nitrogen in the other choices is greater than +2.

23. Which of the following pairs of substances would make a good buffer solution?

(A) $HC_2H_3O_2$ (aq) and $NaC_2H_3O_2$ (aq)
(B) H_2SO_4 (aq) and LiOH (aq)
(C) HCl (aq) and KCl (aq)
(D) HF (aq) and NH_3 (aq)

23. **A** A buffer is made up of either a weak acid and its salt or a weak base and its salt. Option (B) has a strong acid and strong base, option (C) has a strong acid and its salt, and option (D) has a weak acid and a weak base.

QUESTIONS	EXPLANATIONS

Questions 24-27 refer to the following.

Inside a calorimeter, 100.0 mL of 1.0 M hydrocyanic acid (HCN), a weak acid, and 100.0 mL of 0.50 M sodium hydroxide are mixed. The temperature of the mixture rises from 21.5 °C to 28.5 °C. The specific heat of the mixture is approximately 4.2 J/g°C, and the density is identical to that of water.

24. Identify the correct net ionic equation for the reaction that takes place.

    (A)   HCN (aq) + OH$^-$ (aq) ↔ CN$^-$ (aq) + H$_2$O (l)
    (B)   HCN (aq) + NaOH (aq) ↔ NaCN (aq) + H$_2$O (l)
    (C)   H$^+$ (aq) + OH$^-$ (aq) ↔ H$_2$O (l)
    (D)   H$^+$ (aq) + CN$^-$ (aq) + Na$^+$ (aq) + OH$^-$ (aq) ↔ H$_2$O (l) + CN$^-$ (aq) + Na$^+$ (aq)

24. **A** HCN will lose its proton to the hydroxide, creating a conjugate base and water.

25. What is the approximate enthalpy change for the reaction?

    (A)   1.5 kJ
    (B)   2.9 kJ
    (C)   5.9 kJ
    (D)   11.8 kJ

25. **C** $q = mc\Delta T$. The mass of the mixture is 200.0 g; this is from the volume being 200.0 mL and the density of the mixture being 1.0 g/mL. ΔT is 7.0 °C and c is 4.2 J/g°C. So:

$$q = (200.0 \text{ g})(4.2 \text{ J/g°C})(7.0 \text{ °C}) = 5880 \text{ J}.$$

26. At ΔT increases, what happens to the equilibrium constant and why?

    (A)   The equilibrium constant increases because more products are created.
    (B)   The equilibrium constant increases because the rate of the forward reaction increases.
    (C)   The equilibrium constant decreases because the equilibrium shifts to the left.
    (D)   The value for the equilibrium constant is unaffected by temperature and will not change.

26. **C** This is an exothermic reaction; therefore heat is generated as a product. An increase in temperature thus causes a shift to the left, decreasing the numerator and increasing the denominator in the equilibrium expression. This decreases the overall value of K.

27. If the experiment is repeated with 200.0 mL of 1.0 M HCN and 100. mL of 0.50 M NaOH, what would happen to the values for ΔT and the ΔH?

	ΔT	ΔH_{rxn}
(A)	Increase	Increase
(B)	Stay the same	Stay the same
(C)	Decrease	Stay the same
(D)	Stay the same	Increase

27. **C** The NaOH is limiting (0.050 mol vs. 0.100 mol in the original reaction), and adding even more excess HCN will not change the amount of HCN that acutally reacts with OH⁻. So, the value for H stays the same. However, the overall mixture will have a greater mass (300.0 g), which means the temperature change will not be as large.

28. The following diagrams show the Lewis structures of four different molecules. Which molecule would travel the farthest in a paper chromatography experiment using a polar solvent?

28. **A** In a polar solvent, polar molecules will be the most soluble (like dissolves like). Of the four options, methanol and acetone would both have dipoles, but those of methanol would be significantly stronger due to the H-bonding

$$H-\overset{\overset{\displaystyle H}{|}}{\underset{\underset{\displaystyle H}{|}}{C}}-\overset{..}{\underset{..}{O}}-H$$

Methanol

$$H-\overset{\overset{\displaystyle H}{|}}{\underset{\underset{\displaystyle H}{|}}{C}}-\overset{\overset{\displaystyle H}{|}}{\underset{\underset{\displaystyle H}{|}}{C}}-\overset{\overset{\displaystyle H}{|}}{\underset{\underset{\displaystyle H}{|}}{C}}-\overset{\overset{\displaystyle H}{|}}{\underset{\underset{\displaystyle H}{|}}{C}}-\overset{\overset{\displaystyle H}{|}}{\underset{\underset{\displaystyle H}{|}}{C}}-H$$

Pentane

$$H-\overset{\overset{\displaystyle H}{|}}{\underset{\underset{\displaystyle H}{|}}{C}}-\overset{\overset{\displaystyle O}{||}}{C}-\overset{\overset{\displaystyle H}{|}}{\underset{\underset{\displaystyle H}{|}}{C}}-H$$

Acetone

$$H-\overset{\overset{\displaystyle H}{|}}{\underset{\underset{\displaystyle H}{|}}{C}}-O-\overset{\overset{\displaystyle H}{|}}{\underset{\underset{\displaystyle H}{|}}{C}}-H$$

Ether

(A) Methanol
(B) Pentane
(C) Acetone
(D) Ether

29. The first ionization energy for a neutral atom of chlorine is 1.25 MJ/mol and the first ionization energy for a neutral atom of argon is 1.52 MJ/mol. How would the first ionization energy value for a neutral atom of potassum compare to those values?

(A) It would be greater than both because potassium carries a greater nuclear charge then either chlorine or argon.

(B) It would be greater than both because the size of a potassium atom is smaller than an atom of either chlorine or argon.

(C) It would be less than both because there are more electrons in potassium, meaning they repel each other more effectively and less energy is needed to remove one.

(D) It would be less than both because a valence electron of potassium is further from the nucleus than one of either chlorine or argon.

29. **D** Potassium's first valence electron is in the fourth energy level, but both chlorine and argon's first valence electron is in the third energy level.

30. Which net ionic equation below represents a possible reaction that takes place when a strip of magnesium metal is oxidized by a solution of chromium (III) nitrate?

(A) $Mg\ (s) + Cr(NO_3)_3\ (aq) \rightarrow Mg^{2+}(aq) + Cr^{3+}\ (aq) + 3\ NO_3^-\ (aq)$

(B) $3\ Mg\ (s) + 2\ Cr^{3+} \rightarrow 3\ Mg^{2+} + 2\ Cr\ (s)$

(C) $Mg\ (s) + Cr^{3+} \rightarrow Mg^{2+} + Cr\ (s)$

(D) $3\ Mg\ (s) + 2\ Cr(NO_3)_3\ (aq) \rightarrow 3\ Mg^{2+}\ (aq) + 2\ Cr\ (s) +\ NO_3^-\ (aq)$

30. **B** Cr^{3+} needs to gain three electrons to reduce into Cr (s), while Mg (s) loses 2 electrons to oxidize into Mg^{2+}. To balance the electrons, the reduction half-reaction must be multiplied by two and the oxidation half-reaction must be multiplied by three in order to have six electrons on each side. The nitrate ion is a spectator and would not appear in the net ionic equation.

31. $PCl_3\ (g) + Cl_2\ (g) \leftrightarrow PCl_5\ (g) \quad \Delta H = -92.5$ kJ/mol

In which of the following ways could the reaction above be manipulated to create more product?

(A) Decreasing the concentration of PCl_3

(B) Increasing the pressure

(C) Increasing the temperature

(D) None of the above

31. **B** Increasing the pressure in a equilibrium with any gas molecules causes a shift to the side with less gas molecules—in this case, the product.

QUESTIONS	EXPLANATIONS

32. A pure solid substance is heated strongly. It first melts into a liquid, then boils and becomes a gas. Which of the following heating curves correctly shows the relationship between temperature and heat added?

32. **A** During a phase change, all of the energy added goes into break intermolecular forces holding the molecules together. During this time, the speed of the molecules (and thus the temperature) does not rise.

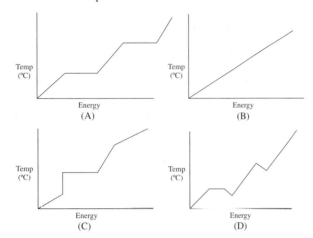

33. Consider the following reaction showing photosynthesis:

$$6 \text{ CO}_2 (g) + 6 \text{ H}_2\text{O} (l) \rightarrow \text{C}_6\text{H}_{12}\text{O}_6 (s) + 6 \text{ O}_2 (g)$$
$$\text{H} = + 2800 \text{ kJ/mol}$$

Which of the following is true regarding the thermal energy in this system?

(A) It is transferred from the surroundings to the reaction.
(B) It is transferred from the reaction to the surroundings.
(C) It is transferred from the reactants to the products.
(D) It is transaction from the products to the reactants.

33. **A** In an endothermic reaction, heat is transferred into the reaction system.

34. $$\text{SO}_2\text{Cl}_2 \rightarrow \text{SO}_2 (g) + \text{Cl}_2 (g)$$

At 600 K, SO_2Cl_2 will decompose to form sulfur dioxide and chlorine gas via the above equation. If the reaction is found to be first order overall, which of the following will cause an increase in the half life of SO_2Cl_2?

(A) Increasing the initial concentration of SO_2Cl_2.
(B) Increasing the temperature at which the reaction occurs.
(C) Decreasing the overall pressure in the container.
(D) None of these will increase the half life.

34. **D** Half-life is independent of all external conditions.

QUESTIONS	EXPLANATIONS

Questions 35-38 refer to the following reaction.

$$2\, SO_2\,(g) + O_2\,(g) \rightarrow 2\, SO_3\,(g)$$

4.0 mol of gaseous SO_2 and 6.0 mol of O_2 gas are allowed to react in a sealed container.

35. Which particulate drawing best represents the contents of the flask after the reaction goes to completion?

(A)

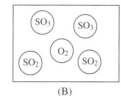

(B)

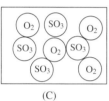

(C)

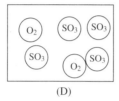

(D)

36. What percentage of the original pressure will the final pressure in the container be equal to?

(A) 67%
(B) 80%
(C) 100%
(D) 133%

37. At a given point in the reaction, all three gases are present at the same temperature. Which gas molecules will have the highest velocity and why?

(A) The O_2 molecules because they are the lightest.
(B) The O_2 molecules because they are the smallest.
(C) The SO_3 molecules because they are products in the reaction.
(D) Molecules of all three gases will have the same speed because they have the same temperature.

35. **C** To determine the number of moles of SO_3 created, stoichiometry must be used.

$$SO_2: \quad 4\ mol\ SO_2 \times \frac{2\ mol\ SO_3}{2\ mol\ SO_2} = 4\ mol\ SO_3$$

$$O_2: \quad 6\ mol\ O_2 \times 2\ mol\ SO_3 = 12\ mol\ SO_3$$

The oxygen is in excess, and only 2.0 mol of it will react. (As every 2 moles SO_2 react with 1 mol of O_2). Thus, 4 mol of SO_3 are created and 4.0 mol of O_2 remain.

36. **B** If 4 mol of SO_3 are created and 4.0 mol of O_2 remain, there are 8.0 mol of gas present after the reaction. Prior to the reaction there were 10.0 mol of gas. If there are 8/10 = 80% as many moles after the reaction, there is also 80% as much pressure.

37. **A** If all gases are at the same temperature, they have the same amount of kinetic energy. Given that $KE = \frac{1}{2}\, mv^2$, if all three gases have the same KE the gas with the least mass must have the highest velocity in order to compensate.

QUESTIONS	EXPLANATIONS

38. Under which of the following conditions would the gases in the container most deviate from ideal conditions and why?

   (A) Low pressures because the gas molecules would be spread far apart.
   (B) High pressures because the gas molecules will be colliding frequently.
   (C) Low temperatures because the intermolecular forces between the gas molecules would increase.
   (D) High temperatures because the gas molecules are moving too fast to interact with each other.

38. **C** One of the assumptions of kinetic molecular theory is that the amount of intermolecular forces between the gas molecules is negligible. If the molecules are moving very slowly, the IMFs between them are more likely to cause deviations from ideal behavior.

39. Four different acids are added to beakers of water, and the following diagrams represent the species present in each solution at equilibrium. Which acid has the highest pH?

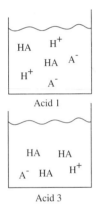

Acid 1

Acid 2

Acid 3

Acid 4

   (A) Acid 1
   (B) Acid 2
   (C) Acid 3
   (D) Acid 4

39. **C** The strength of an acid is dependent on the amount it dissociates in solution. The less dissociation there is, the less hydrogen ions there are present and the weaker the acid is, leading to a higher pH.

40. Which expression below should be used to calculate the mass of copper than can be plated out of a 1.0 M $Cu(NO_3)_2$ solution using a current of 0.75 A for 5.0 minutes?

   (A) $\dfrac{(5.0)(60)(0.75)(63.55)}{(96500)(2)}$
   (B) $\dfrac{(5.0)(60)(63.55)(2)}{(0.75)(96500)}$
   (C) $\dfrac{(5.0)(60)(96500)(0.75)}{(63.55)(2)}$
   (D) $\dfrac{(5.0)(60)(96500)(63.55)}{(0.75)(2)}$

40. **A**

$$5.0 \text{ min} \times \frac{60.0 \text{ s}}{1.0 \text{ min}} \times \frac{0.75 \text{ C}}{1.0 \text{ s}} \times \frac{1 \text{ mol e}^-}{96500 \text{ C}} \times$$

$$\frac{1 \text{ mol Cu}}{2 \text{ mol e}^-} \times \frac{63.55 \text{ g Cu}}{1 \text{ mol Cu}}$$

QUESTIONS	EXPLANATIONS

41. Lewis diagrams for the nitrate and nitrite ions are shown above. Choose the statement that correctly describes the relationship between the two ions in terms of bond length and bond energy.

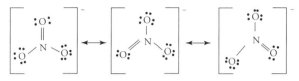

Nitrate

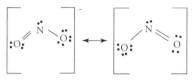

Nitrite

(A) Nitrite has longer and stronger bonds than nitrate.
(B) Nitrite has longer and weaker bonds than nitrate.
(C) Nitrite has shorter and stronger bonds than nitrate.
(D) Nitrite has shorter and weaker bonds than nitrate.

41. **C** Nitrate has a bond order of $(1 + 3/3 = 1.33)$ Nitite has a bond order of $(1 + 2/2 = 1.5)$. A higher bond order means shorter and stronger bonds.

42. Examining data obtained from mass spectrometry supports which of the following?

(A) the common oxidation states of elements
(B) atomic size trends within the periodic table
(C) ionization energy trends within the periodic table
(D) the existence of isotopes

42. **D** Mass spectrometry is used to determine the masses for individual atoms of an element. Through mass spectrometry, it is proven that each element has more than one possible mass.

43. A 2.0 L flask holds 0.40 g of Helium gas. If the helium is evacuated into a larger container while the temperature is held constant, what will the effect on the entropy of the helium be?

(A) It will remain constant as the number of helium molecules does not change.
(B) It will decrease as the gas will be more ordered in the larger flask.
(C) It will decrease because the molecules will collide with the sides of the larger flask less often than they did in the smaller flask.
(D) It will increase as the gas molecules will be more dispersed in the larger flask.

43. **D** Entropy is a measure of a system's disorder. In a larger flask, the gas molecules will spread further apart and become more disordered.

Questions 44-48 refer to the following.

NO$_2$ gas is placed in a sealed, evacuated container and allowed to decompose via the following equation:

$$2\,NO_2\,(g) \leftrightarrow 2\,NO\,(g) + O_2\,(g)$$

The graph below indicates the change in concentration for each species over time.

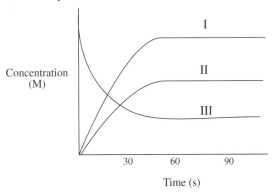

44. Using the numbers of the lines on the graph, identify which line belongs to which species.

	Line I	Line II	Line III
(A)	NO$_2$	NO	O$_2$
(B)	O$_2$	NO$_2$	NO
(C)	NO	O$_2$	NO$_2$
(D)	O$_2$	NO	NO$_2$

44. **C** The only species that is present at $t = 0$ is the NO$_2$, allowing us to identify line III. When identifying line I vs. line II, the NO will be generated twice as quickly as the O$_2$ due to the coefficients, meaning [NO] will increase about twice as quickly as [O$_2$].

45. What is happening to the rate of the forward reaction at $t = 60$ s?

(A) It is increasing.
(B) It is decreasing.
(C) It is remaining constant.
(D) It has stopped.

45. **C** At equilibrium, the concentration of all species in the reaction are remaining constant, which shows up as a flat line on the graph. The rate of both the forward and reverse reactions are constant at equilibrium.

QUESTIONS	EXPLANATIONS
46. At the reaction progresses, what happens to the value of the equilibrium constant K_p if the temperature remains constant? (A) It stays constant. (B) It increases exponentially. (C) It increases linearly. (D) It decreases exponentially.	46. **A** The only factor that can affect the value of the equilibrium constant is temperature. If the temperature does not change, neither does the equilibrium constant.
47. What would happen to the slope of the NO_2 line if additional O_2 were injected into the container? (A) It would increase, then level off. (B) It would decrease, then level off. (C) It would remain constant. (D) It would increase, then decrease.	47. **A** If additional O_2 were injected into the container, the reaction would shift left, increasing the amount of NO_2 present. Eventually, the reaction would reach equilibrium again, meaning the lines would level out.
48. Using the graph, how could you determine the instantaneous rate of disappearance of NO_2 at $t = 30$ s? (A) By determining the area under the graph at $t = 30$ s (B) By taking the slope of a line tangent to the NO_2 curve at $t = 30$ s (C) By using the values at $t = 30$ s and plugging them into the K_p expression (D) By measuring the overall gas pressure in the container at $t = 30$ s	48. **B** To determine the change in concentration at a specific time, we would need the slope of the line at that point. As the line is curved, the only way to do that (without calculus) is to draw a line tangent to the curve at that point and measure its slope.
49. A proposed mechanism for a reaction is as follows: $NO_2 + F_2 \rightarrow NO_2F + F$ Slow step $F + NO_2 \rightarrow NO_2F$ Fast step What is the order of the overall reaction? (A) Zero order (B) First order (C) Second order (D) Third order	49. **C** The overall rate law is always equal to the rate law for slowest elementary step, which can be determined using the coefficients of the reactants. In this case, rate = $k[NO_2][F_2]$. To get the overall order, we add the exponents in the rate law. $1 + 1 = 2$

50. Starting with a stock solution of 18.0 M H_2SO_4, what is the proper procedure to create a 1.00 L of a 3.0 M solution of H_2SO_4 in a volumetric flask?

(A) Add 167 mL of the stock solution to the flask, then fill the flask the rest of the way with distilled water while swirling the solution.

(B) Add 600 mL of the stock solution to the flask, then fill the flask the rest of the way with distilled water while swirling the solution.

(C) Fill the flask partway with water, then add 167 mL of the stock solution, swirling to mix it. Last, fill the flask the rest of the way with distilled water.

(D) Fill the flask partway with water, then add 600 mL of the stock solution, swirling to mix it. Last, fill the flask the rest of the way with distilled water.

50. **C** Use $M_1V_1 = M_2V_2$ to determine the necessary volume of stock solution.

$$(18.0 \text{ M})V_1 = (3.0 \text{ M})(1.0 \text{ L}) \quad V_1 = 0.167 \text{ L} = 167 \text{ mL}$$

When creating solutions with acid, you always add some water first, as the process is extremely exothermic and the water will absorb the generated heat.

51. The reaction shown in the above diagram is accompanied by a large increase in temperature. If all molecules shown are in their gaseous state, which statement accurately describes the reaction?

(A) It is an exothermic reaction in which entropy increases.

(B) It is an exothermic reaction in which entropy decreases.

(C) It is an endothermic reaction in which entropy increases.

(D) It is an endothermic reaction in which entropy decreases.

51. **A** The temperature increase is indicative of energy being released, meaning the reaction is exothermic. The entropy (disorder) of the system is increasing as it moves from three gas molecules to five.

QUESTIONS	EXPLANATIONS

52. A student mixes equimolar amounts of KOH and $Cu(NO_3)_2$ in a beaker. Which of the following particulate diagrams correctly shows all species present after the reaction occurs?

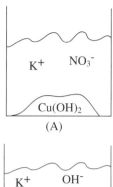

(A)

(B)

(C)

(D)

52. D The overall reaction (including spectator ions) is: $2\ OH^- + Cu^{2+} \rightarrow Cu(OH)_2(s)$. Both the K^+ and the NO_3^- are spectator ions which are present in solution both before and after the reaction. Additionally, if equimolar amounts of the two reactants are initially present, the OH^- will run out before the Cu^{2+}, meaning that some of those ions will also be present in the final solution.

Questions 53-55 refer to the following.

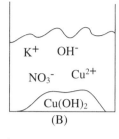

Methanol

Propane

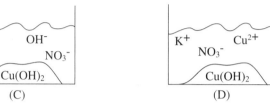

Ethene

Ethanal

53. Based on the strength of the intermolecular forces in each substance, estimate from greatest to smallest the vapor pressures of each substance in liquid state at the same temperature.

(A) propane > ethanal > ethene > methanol
(B) ethene > propane > ethanal > methanol
(C) ethanal > methanol > ethene > propane
(D) methanol > ethanal > propane > ethene

53. B Vapor pressure is dependent on intermolecular forces. The weaker the IMFs are, the easier it is for molecules to escape from the surface of the liquid. To begin, polar molecules have stronger IMFs than nonpolar molecules. Methanol and ethanal are both poplar, but methanol has hydrogen bonding meaning it has stronger IMFs (and thus a lower vapor pressure) than ethanal. Ethene and propane are both nonpolar, but propane is larger meaning it is more polarizable than ethene and thus has stronger IMFs and lower vapor pressure.

QUESTIONS	EXPLANATIONS
54. When in liquid state, which two substances are most likely to be miscible with water? (A) propane and ethene (B) methanol and propane (C) ethene and ethanal (D) methanol and ethanal	54. **D** Water is polar and using "like dissolves like", we know that only polar solvents will be able to fully mix with it to create a homogenous solution.
55. Between propane and ethene, which will likely have the higher boiling point and why? (A) Propane, because it has a greater molar mass (B) Propane, because it has a more polarizable electron cloud (C) Ethene, because of the double bond (D) Ethene, because it is smaller in size	55. **B** Both are nonpolar, but propane has a lot more electrons and thus is more polarizable than ethene.
56. $4 NH_3 (g) + 5 O_2 (g) \rightarrow 4 NO (g) + 6 H_2O (g)$ The above reaction will experience a rate increase by the addition of a cataylst such as platinum. Which of the following best explains why? (A) The catalyst increases the overall frequency of collisions in the reactant molecules. (B) The catalyst increases the frequency of collisions that occur at the proper orientation in the reactant molecules. (C) The catalyst introduces a new reaction mechanism for the reaction. (D) The catalyst increases the activation energy for the reaction.	56. **C** Catalysts work by creating a new reaction pathway with a lower activation energy than the original pathway.

QUESTIONS	EXPLANATIONS

57. The graph below shows the amount of potential energy between two hydrogen atoms as the distance between them changes. At which point in the graph would a molecule of H_2 be the most stable?

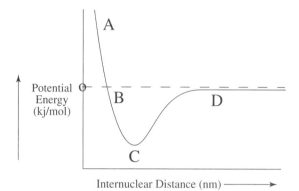

(A) Point A
(B) Point B
(C) Point C
(D) Point D

57. **C** The molecule would be the most stable when it has the largest attractive potential energy, which is represented by a negative sign. While the magnitude of the potential energy may be larger at (A), it is repulsive at that point- the nuclei are too close together.

58. $N_2(g) + O_2(g) + Cl_2(g) \leftrightarrow 2NOCl(g)$ $\Delta G = 132.6$ kJ/mol

For the equilibrium above, what would happen to the value of ΔG if the concentration of N_2 were to increase and why?

(A) It would increase as the reaction would become more thermodynamically favored.
(B) It would increase as the reaction would shift right and create more products.
(C) It would decrease because there are more reactants present.
(D) It would stay the same because the value of K_{eq} would not change.

58. **D** Adding (or removing) any species in an equilibrium reaction does not change the equilibrium constant and also does not change the magnitude of the gibbs free energy.

QUESTIONS	EXPLANATIONS

59. $C\,(s) + 2\,S\,(s) \rightarrow CS_2\,(l)$ $\Delta H = +92.0\ \text{kJ/mol}$

Which of the following energy level diagrams gives an accurate representation of the above reaction?

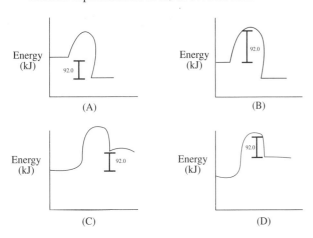

(A)   (B)

(C)   (D)

59. **C**   A positive H means the reaction is endothermic, so the products have more bond energy than the reactants. The different between the energy levels of the products and reactants is equal to ΔH (the different between the energy level of the reactants and the top of the hump is the value for the activation energy).

60. Two alloys are shown in the diagrams above—bronze, and steel. Which of the following correctly describes the malleability of both alloys compared to their primary metals?

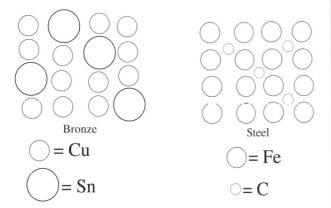

Bronze   Steel

⃝ = Cu   ⃝ = Fe

⃝ = Sn   ○ = C

(A) Bronze's malleability would be comparable to that of copper, but steel's malleability would be significantly lower than that of iron.
(B) Bronze's malleability would be significantly higher than that of copper, but steel's malleability would be comparable to that of iron.
(C) Both bronze and steel would have malleability values similar to those of their primary metals.
(D) Both bronze and steel would have malleability values greater than those of their primary metals.

60. **A**   Bronze is a substitutional alloy, in which some of the copper atoms are repalced with those of tin. As the tin and copper atoms are both the same size, the overall malleability of the alloy would not change significantly. Steel is an interstitial alloy, in which carbon atoms are places in between iron atoms. This adds significant mass to the lattice and reduces the malleability of the alloy.

Section II—Constructed Response

1. A student is tasked with determining the identity of an unknown carbonate compound with a mass of 1.89 g. The compound is first placed in water, where it dissolves completely. The K_{sp} value for several carbonate-containing compounds are given below.

Compound	K_{sp}
Lithium Carbonate	8.15×10^{-4}
Nickel (II) Carbonate	1.42×10^{-7}
Strontium Carbonate	5.60×10^{-10}

(a) In order to precipitate the maximum amount of the carbonate ions from solution, which of the following should be added to the carbonate solution: $LiNO_3$, $Ni(NO_3)_2$, or $Sr(NO_3)_2$? Justify your answer.

(a) The student should use the strontium nitrate. Using it would create strontium carbonate, which has the lowest K_{sp} value. That means it is the least soluble carbonate compound of the three, and will precipitate the most possible carbonate ions out of solution.

(b) For the carbonate compound that contains the cation chosen in part (a), determine the concentration of each ion of that compound in solution at equilibrium.

(b) $SrCO_3\ (s) \leftrightarrow Sr^{2+}(aq) + CO_3^{2-}(aq)$

$K_{sp} = [Sr^{2+}][CO_3^{2-}]$

$5.60 \times 10^{-10} = (x)(x)$

$x = 2.37 \times 10^{-5}\ M = [Sr^{2+}] = [CO_3^{2-}]$

(c) When mixing the solution, should the student ensure the carbonate solution or the nitrate solution is in excess? Justify your answer.

(c) The ntirate solution should be in excess. In order to create the maximum amount of precipitate, enough strontium ions need to be added to react with all of the carbonate ions originally in solution. Having excess strontium ions in solution after the precipitate forms will not affect the calculated mass of the carbonate in the original sample.

QUESTIONS	EXPLANATIONS

After titrating sufficient solution to precipitate out all of the carbonate ions, the student filters the solution before placing it in a crucible and heating it to drive off the water. After several heatings, he final mass of the precipitate remains constant and is determined to be 2.02 g.

(d)

   (i) Determine the number of moles of precipitate.

(i) $2.02 \text{ g SrCO}_3 \times \dfrac{1 \text{ mol SrCO}_3}{147.63 \text{ g SrCO}_3}$

$= 1.34 \times 10^{-2} \text{ mol SrCO}_3$

   (ii) Determine the mass of carbonate present in the precipitate.

(ii) $1.34 \times 10^{-2} \text{ mol SrCO}_3 \times$

$\dfrac{1 \text{ mol CO}_3^{2-}}{1 \text{ mol SrCO}_3} \times \dfrac{60.01 \text{ g CO}_3^{2-}}{1 \text{ mol CO}_3^{2-}} = 0.821 \text{ g CO}_3^{2-}$

(e) Determine the percent, by mass, of carbonate in the original sample.

(e) $\dfrac{0.82 \text{ g}}{1.39 \text{ g}} \times 100 = 43.4\% \text{ CO}_3^{2-}$

(f) Is the original compound most likely lithium carbonate, sodium carbonate, or potassium carbonate? Justify your answer.

(f)   The mass percent of carbonate in each compound must be compared to the experimental determined mass percent of carbonate in the sample.

$\text{Li}_2\text{CO}_3: \dfrac{60.01}{73.89} \times 100 = 81.2\%$

$\text{Na}_2\text{CO}_3: \dfrac{60.01}{105.99} \times 100 = 56.6\%$

$\text{K}_2\text{CO}_3: \dfrac{60.01}{138.21} \times 100 = 45.4\%$

The compound is most likely potassium carbonate.

2. The **unbalanced** reaction between potassium permanganate and acidified iron (II) sulfate is a redox reaction that proceeds as follows:

$$H^+ (aq) + Fe^{2+} (aq) + MnO_4^- (aq) \rightarrow$$
$$Mn^{2+} (aq) + Fe^{3+} (aq) + H_2O (l)$$

(a) Provide the equations for both half reactions that occur below:

(i) Oxidation half reaction:

(i) Oxidation: $Fe^{2+} (aq) \rightarrow Fe^{3+} (aq) + e^-$

(ii) Reduction half reaction:

(ii) Reduction: $5e^- + 8H^+ (aq) + MnO_4^- (aq) \rightarrow$
$$Mn^{2+} (aq) + 4 H_2O (l)$$

(b) What is the balanced net ionic equation?

(b) The oxidation reaction must be multiplied by a factor of 5 in order for the electrons to balance out. So:

$$5 Fe^{2+} (aq) + 8 H^+ (aq) + MnO_4^- (aq) \rightarrow 5 Fe^{3+} (aq) +$$
$$Mn^{2+} (aq) + 4 H_2O (l)$$

A solution of 0.150 M potassium permanganate, a dark purple solution, is placed in a buret before being titrated into a flask containing 50.00 mL of iron (II) sulfate solution of unknown concentration. The following data describes the colors of the various ions in solution:

Ion	Color in solution	Ion	Color in Solution
H^+	Colorless	Fe^{3+}	Yellow
Fe^{2+}	Pale Green	K^+	Colorless
MnO_4^-	Dark Purple	SO_4^{2-}	Colorless
Mn^{2+}	Colorless		

(c) Describe the color of the solution in the flask at the following points:

(i) Before titration begins

(i) The only ions present in the flask are Fe^{2+}, SO_4^{2-}, and H^+. The latter two are colorless, so the solution would be pale green.

QUESTIONS	EXPLANATIONS
(ii) During titration prior to the endpoint	(ii) The MnO_4^- is reduced to Mn^{2+} upon entering the flask, and the Fe^{2+} ions are oxidized into Fe^{3+} ions. The solution would become less green and more yellow and more Fe^{3+} ions are formed, as all other ions present are colorless.
(iii) At the endpoint of the titration	(iii) After the Fe^{2+} ions have all been oxidized, there is nothing left to donate electrons to the MnO_4^- ions. Therefore, they will no longer be reduced upon entering the flask, and the solution will take on a light purplish/yellow hue due to the mixture of MnO_4^- and Fe^{3+} ions.

(d) (i) If 15.55 mL of permanganate are added to reach the endpoint, what is the initial concentration of the iron (II) sulfate?

(i) First the moles of permanganate added must be calculated:

$$0.150 \text{ M} = \frac{n}{0.0155 \text{ L}} \quad n = 2.33 \times 10^{-3} \text{ mol } MnO_4^-$$

Then the moles of iron (II) can be determined via stoichiometry:

$$2.33 \times 10^{-3} \text{ mol } MnO_4^- \times \frac{5 \text{ mol Fe}^{2+}}{1 \text{ mol MnO}_4} = 0.0117 \text{ mol Fe}^{2+}$$

Finally the concentration of the $FeSO_4$ can be determined:

$$\text{Molarity} = \frac{0.0117 \text{ mol}}{0.050 \text{ L}} = 0.234 \text{ M}$$

(ii) The actual concentration of the $FeSO_4$ is 0.250 M. Calculate the percent error.

(ii)

$$\text{Percent error} = \frac{\text{actual value} - \text{experimental value}}{\text{actual value}} \times 100$$

$$\text{Percent error} = \frac{0.250 - 0.234}{0.250} \times 100 = 6.40\% \text{ error}$$

QUESTIONS	EXPLANATIONS

(e) Could the following errors have led to the experimental result deviating in the direction that it did? You must justify your answers quantitatively.

   (i) 55.0 mL of $FeSO_4$ was added to the flask prior to titration instead of 50.0 mL.

(i) If the volume of $FeSO_4$ was artificially low in the calculations, that would lead to the experimental value for the concentration of $FeSO_4$ being artificially high. As the calculated value for the concentration of $FeSO_4$ was too low, this error source is not supported by data.

   (ii) The concentration of the potassium permanganate was actually 0.160 M instead of 0.150 M.

(ii) If the molarity of the permanganate was artificially low in the calculations, the moles of permanganate, and by extension, the moles of Fe^{2+} would also be artificially low. This would lead to an artificially low value for the concentration of $FeSO_4$. This matches with the experimental results and is thus supported by the data.

3. Propane gas can be burned in air via the following reaction:

$$2\ NH_3\ (g) + 3\ N_2O\ (g) \rightarrow 4\ N_2\ (g) + 3\ H_2O\ (g)$$

The standard entropy of formation values for the varying substances are listed in the table below.

Substance	S° (J/molK)
NH_3 (g)	193
N_2O (g)	220
N_2 (g)	192
H_2O (g)	189

(a) Calculate the entropy value for the overall reaction

(a) $\Delta S = \Delta S_{products} - \Delta S_{reactants}$

$\Delta S = [3(H_2O) + 4(N_2)] - [2(NH_3) + 3(N_2O)]$

$\Delta S = [3(189) + 4(192)] - [2(193) + 3(220)]$

$\Delta S = 1335 - 1046$

$\Delta S = 289$ J/mol*K

QUESTIONS

Several bond enthalpies are listed in the table below.

Bond	Enthalpy (kJ/mol)	Bond	Enthalpy (kJ/mol)
N–H	388	N=N	409
N–O	210	N≡N	941
N=O	630	O–H	463

(b) Calculate the enthalpy value for the overall reaction

EXPLANATIONS

(b) To determine the amount of enthalpy change, Lewis structures of all of the species should be drawn.

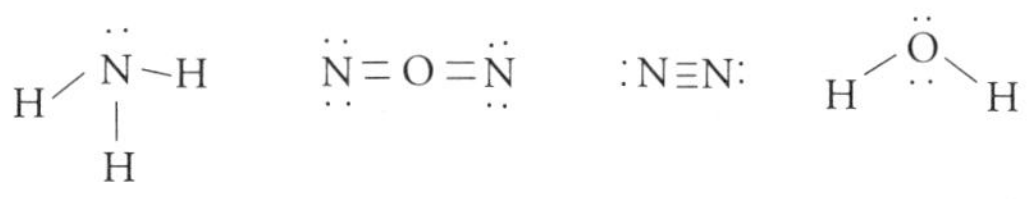

ΔH = Bonds broken (reactants) – bonds formed (products)

ΔH = [6(N-H) + 6(N=O)] – [4(N N) + 6(O-H)]

ΔH = [6(388) + 6(630)] – [4(941) + 6(463)]

ΔH = 6108 – 6542

ΔH = –434 kJ/mol

(c) Is this reaction thermodynamically favored at 25°C? Justify your answer.

(c) We will use Gibbs free energy equation here. Keep in mind the units have to match. In this solution, the units for entropy have been converted to kJ.

$\Delta G = \Delta H - T\Delta S$

ΔG = -434 kJ/mol – (298 K)(0.259 kJ/mol*K)

ΔG = –434 J/mol – 77 kJ/mol

ΔG = –511 kJ/mol

The value for ΔG is negative, therefore the reaction is thermodynamically favored at 25°C.

(d) If 25.00 g of NH_3 reacts with 25.00 g of N_2O:
(i) Will energy be released or absorbed?

(i) The value for H is negative; therefore it is an exothermic reaction and energy will be released.

(ii) What is the magnitude of the energy change?

(ii) To determine how much energy is released, the limiting reagent must be determined.

$$25.00 \text{ g } NH_3 \times \frac{1 \text{ mol } NH_3}{17.04 \text{ g } NH_3} \times \frac{1 \text{ mol}_{rxn}}{2 \text{ mol } NH_3} \times \frac{-511 \text{ kJ}}{1 \text{ mol}_{rxn}}$$

$$= -374.9 \text{ kJ}$$

$$25.00 \text{ g } N_2O \times \frac{1 \text{ mol } N_2O}{44.02 \text{ g } N_2O} \times \frac{1 \text{ mol}_{rxn}}{3 \text{ mol } N_2O} \times \frac{-511 \text{ kJ}}{1 \text{ mol}_{rxn}}$$

$$= -96.74 \text{ kJ}$$

As the N_2O would produce less energy, it would run out first and is thus limiting. The answer is thus −96.74 kJ.

(e) On the reaction coordinates below, draw a line showing the progression of this reaction. Label both ΔH and E_a on the graph.

Potential
Energy
(kJ/mol)

Reaction Progress

(e)

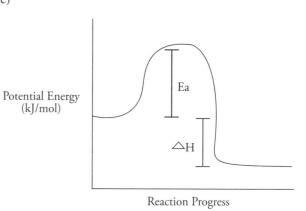

Potential Energy
(kJ/mol)

Ea

△H

Reaction Progress

4. The acetyl ion has a formula of $C_2H_3O^-$ and two possible Lewis electron-dot diagram representations:

(a) Using formal charge, determine which structure is the most likely correct and circle it.

(a) For this formal charge calculation, the H atoms are left out as they are identically bonded/drawn in both structures

	C	C	O		C	C	O
Valence	4	4	6		4	4	6
Assigned	−4	4	7		−4	5	6
Formal Charge	0	0	−1		0	−1	0

As oxygen is more electronegative than carbon, an oxygen atom is more likely to have the negative formal charge than a carbon atom. The left-hand structure is most likely correct.

(b) For Carbon atom "x" in the structure you chose:
(i) What is the hybridization around the atom?

(i) There are three charge groups around the carbon atom, so the hybridization is sp^2

(ii) How many sigma and pi bonds has the atom formed?

(ii) Single bonds consist of sigma bonds, and double bonds consist of one sigma and one pi bond.

There are a total of three sigma bonds and one pi bound around the carbon atom.

QUESTIONS	EXPLANATIONS
(c) A hydrogen ion attaches itself to the the acetyl ion, creating C_2H_4O. Draw the Lewis diagram of the new molecule.	(c) The hydrogen ion will attach to the negatively-charged oxygen.

$$
\begin{array}{ccc}
H & & H \\
| & & | \\
C & = & C - O - H \\
| & & \\
H & &
\end{array}
$$

5. A student tests the conductivity of three different acid samples, each with a concentration of 0.5 M and a volume of 100.0 mL. The conductivity was recorded in microSiemens per centimeter in the table below:

Sample	Conductivity (µS/cm)
1	26820
2	8655
3	35120

(a) The three acids are known to be HCl, H_2SO_4, and H_3PO_4. Identify which sample is which acid. Justify your answer.

(a) H_3PO_4 is the only weak acid, meaning it will not dissociate completely in solution. Therefore, it is acid 2. H_2SO_4 and HCl are both strong acids that will dissociate completely, but H_2SO_4 is diprotic, meaning there will be more H^+ ions in solution and thus a higher conductivity. Therefore, H_2SO_4 is acid 3, and HCl is acid 1.

(b) The HCl solution is then titrated with a 1.50 M solution of the weak base methylamine, CH_3NH_2. ($K_b = 4.38 \times 10^{-4}$)
(i) Write out the net ionic equation for this reaction.

(i) $H^+ + CH_3NH_2 \leftrightarrow CH_3NH_3^+$

QUESTIONS	EXPLANATIONS
(ii) Determine the pH of the solution after 20.0 mL of methylamine has been added.	(ii) First, the number of moles of hydrogen ions and methylamine need to be determined.

(ii) First, the number of moles of hydrogen ions and methylamine need to be determined.

H^+: $0.10 \text{ M} = \dfrac{n}{0.020 \text{ L}}$ $\qquad$ n = 0.0020 mol

CH_3NH_2: $0.150 \text{ M} = \dfrac{n}{0.020 \text{ L}}$ $\qquad$ n = 0.0030 mol

	H^+	+	CH_3NH_2	$\leftrightarrow$	$CH_3NH_3^+$
I	0.0020		0.0030		0
C	−0.0020		−0.0020		+0.0020
E	0		0.0010		0.0020

The new volume of the solution is 40.0 mL, which we used to calculated the concentrations of the ions in solution at equilibrium:

$[CH_3NH_2] = \dfrac{0.0010 \text{ mol}}{0.040 \text{ L}} = 0.025 \text{ M}$

$[CH_3NH_3^+] = \dfrac{0.0020 \text{ mol}}{0.040 \text{ L}} = 0.050 \text{ M}$

Finally, using Henderson-Haselbalch:

$pOH = pK_b + \log\dfrac{\left[CH_3NH_3^+\right]}{\left[CH_3NH_2\right]}$

$pOH = -\log(4.38 \times 10^{-4}) + \log\dfrac{0.050}{0.025}$

$pOH = 3.36 + 0.30 = 3.66$

$pH + pOH = 14$

$pH + 3.66 = 14$

$pH = 10.34$

6. The photoelectron spectrum of an element is given below:

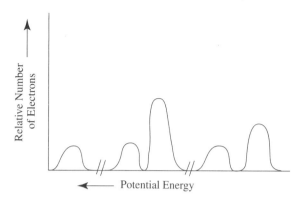

(a) Using your knowledge of atomic structure, explain the following:

(i) The reason for the three discrete areas of ionization energies

(i) Each discrete area of ionization energy represents a different energy level of the electrons. The closer the electrons are to the nucleus, the more Coulombic potential energy they will have, and thus, more ionization energy will be required to remove the electrons. Sulfur has electrons present at three different energy levels, thus, there are three different areas for the peaks

(ii) The justification for there being a total of five peaks

(ii) Within each energy level (except for the first), there are subshells which are not the exact same distance from the nucleus. Both energy levels 2 and 3 have s and p subshells, and while the electrons in those shells will have similar ionization energy values, they will not be identical. Thus, the five peaks represent $1s$, $2s$, $2p$, $3s$, and $3p$.

(iii) The relative heights of the peaks when compared to each other

(iii) The heights of the peaks represents the ratio of electrons present in each of them. All three s peaks are exactly one third the height of the $2p$ peak, meaning there are three times more electrons in $2p$ than in any of the s subshells. The $3p$ peak is only twice as high as the s peaks, therefore, there are twice as many electrons in $3p$ than in any of the s subshells.

(b) Identify the element this spectra most likely belongs to and write out its full electron configuration.

(b) This PES belongs to sulfur: $1s^2 2s^2 2p^6 3s^2 3p^4$

QUESTIONS	EXPLANATIONS

7. Hydrazine (N_2H_4) can be produced commercially via the Raschig process. The following is a proposed mechanism:

Step 1: $NH_3\ (aq) + OCl^-\ (aq) \rightarrow NH_2Cl\ (aq) + OH^-$
Step 2: $NH_2Cl\ (aq) + NH_3\ (aq) \rightarrow N_2H_5^+\ (aq) + Cl^-$
Step 3: $N_2H_5^+\ (aq) + OH^- \rightarrow N_2H_4\ (aq) + H_2O\ (l)$

(a)

(i) What is the equation for the overall reaction?

(i) To determine the net equation, all three equations must be added together, and species that appear on both sides of the arrow can be eliminated.

$NH_3\ (aq) + OCl^-\ (aq) + \cancel{NH_2Cl\ (aq)} + NH_3\ (aq) + \cancel{N_2H_5^+\ (aq)} + \cancel{OH^-} \rightarrow \cancel{NH_2Cl\ (aq)} + \cancel{OH^-} + \cancel{N_2H_5^+\ (aq)} + Cl^- + N_2H_4\ (aq) + H_2O\ (l)$

$2\ NH_3\ (aq) + OCl^-\ (aq) \rightarrow N_2H_4\ (aq) + H_2O\ (l) + Cl^-$

(ii) Identify any catalysts or intermediates from the reaction mechanism.

(ii) There are no catalysts present, but NH_2Cl, $N_2H_5^+$, and OH^- are all intermediates in the process.

The rate law for the reaction is determined to be rate = $k[NH_3][OCl^-]$

(b)

(i) Which elementary step is the slowest one? Justify your answer.

(i) The overall rate law will match the rate law of the slowest step. The rate law of an elementary step can be determined by the reactants present, and in this case, the rate law for Step 1 matches the overall rate law. Therefore, Step 1 is the slowest step.

(ii) If the reaction is measured over the course of several minutes, what would the units on the rate constant be?

(ii) Using unit analysis:

rate = $k[NH_3][OCl^-]$
M/min = $k(M)(M)$
$k = M^{-1}min^{-1}$

Chapter 12
Practice Test 2

AP® Chemistry Exam

DO NOT OPEN THIS BOOKLET UNTIL YOU ARE TOLD TO DO SO.

At a Glance

Total Time
1 hour and 30 minutes
Number of Questions
60
Percent of Total Grade
50%
Writing Instrument
Pencil required

Instructions

Section I of this examination contains 60 multiple-choice questions. Fill in only the ovals for numbers 1 through 60 on your answer sheet.

CALCULATORS MAY NOT BE USED IN THIS PART OF THE EXAMINATION.

Indicate all of your answers to the multiple-choice questions on the answer sheet. No credit will be given for anything written in this exam booklet, but you may use the booklet for notes or scratch work. After you have decided which of the suggested answers is best, completely fill in the corresponding oval on the answer sheet. Give only one answer to each question. If you change an answer, be sure that the previous mark is erased completely. Here is a sample question and answer.

Sample Question   Sample Answer

Chicago is a   Ⓐ ● Ⓒ Ⓓ
(A) state
(B) city
(C) country
(D) continent

Use your time effectively, working as quickly as you can without losing accuracy. Do not spend too much time on any one question. Go on to other questions and come back to the ones you have not answered if you have time. It is not expected that everyone will know the answers to all the multiple-choice questions.

About Guessing

Many candidates wonder whether or not to guess the answers to questions about which they are not certain. Multiple choice scores are based on the number of questions answered correctly. Points are not deducted for incorrect answers, and no points are awarded for unanswered questions. Because points are not deducted for incorrect answers, you are encouraged to answer all multiple-choice questions. On any questions you do not know the answer to, you should eliminate as many choices as you can, and then select the best answer among the remaining choices.

Disclaimer

This test is an approximation of the test that you will take. For up-to-date information, please remember to check the AP Central website.

GO ON TO THE NEXT PAGE.

CHEMISTRY
SECTION I

Time—1 hour and 30 minutes

Material in the following table may be useful in answering the questions in this section of the examination.

PERIODIC CHART OF THE ELEMENTS

1 H 1.0																		2 He 4.0
3 Li 6.9	4 Be 9.0											5 B 10.8	6 C 12.0	7 N 14.0	8 O 16.0	9 F 19.0	10 Ne 20.2	
11 Na 23.0	12 Mg 24.3											13 Al 27.0	14 Si 28.1	15 P 31.0	16 S 32.1	17 Cl 35.5	18 Ar 39.9	
19 K 39.1	20 Ca 40.1	21 Sc 45.0	22 Ti 47.9	23 V 50.9	24 Cr 52.0	25 Mn 54.9	26 Fe 55.8	27 Co 58.9	28 Ni 58.7	29 Cu 63.5	30 Zn 65.4	31 Ga 69.7	32 Ge 72.6	33 As 74.9	34 Se 79.0	35 Br 79.9	36 Kr 83.8	
37 Rb 85.5	38 Sr 87.6	39 Y 88.9	40 Zr 91.2	41 Nb 92.9	42 Mo 95.9	43 Tc (98)	44 Ru 101.1	45 Rh 102.9	46 Pd 106.4	47 Ag 107.9	48 Cd 112.4	49 In 114.8	50 Sn 118.7	51 Sb 121.8	52 Te 127.6	53 I 126.9	54 Xe 131.3	
55 Cs 132.9	56 Ba 137.3	57 *La 138.9	72 Hf 178.5	73 Ta 180.9	74 W 183.9	75 Re 186.2	76 Os 190.2	77 Ir 192.2	78 Pt 195.1	79 Au 197.0	80 Hg 200.6	81 Tl 204.4	82 Pb 207.2	83 Bi 209.0	84 Po (209)	85 At (210)	86 Rn (222)	
87 Fr (223)	88 Ra 226.0	89 †Ac 227.0																

*Lanthanum Series	58 Ce 140.1	59 Pr 140.9	60 Nd 144.2	61 Pm (145)	62 Sm 150.4	63 Eu 152.0	64 Gd 157.3	65 Tb 158.9	66 Dy 162.5	67 Ho 164.9	68 Er 167.3	69 Tm 168.9	70 Yb 173.0	71 Lu 175.0
†Actinium Series	90 Th 232.0	91 Pa 231.0	92 U 238.0	93 Np 237.0	94 Pu (244)	95 Am (243)	96 Cm (247)	97 Bk (247)	98 Cf (251)	99 Es (252)	100 Fm (258)	101 Md (258)	102 No (259)	103 Lr (260)

DO NOT DETACH FROM BOOK.

GO ON TO THE NEXT PAGE.

1. A compound is made up of entirely silicon and oxygen atoms. If there is 14.00 g of silicon and 32.0 g of oxygen present, what is the empirical formula of the compound?

    (A) SiO_2
    (B) SiO_4
    (C) Si_2O
    (D) Si_2O_3

2.

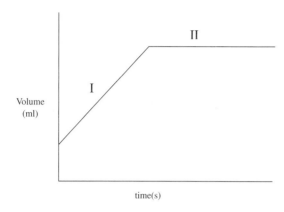

time(s)

The volume of a gas is charted over time, giving the above results. Which of the following options provides a possible explanation of what was happening to the gas during each phase of the graph?

(A) During phase I, the temperature decreased while the pressure increased. During phase II, the temperature was held constant as the pressure decreased.
(B) During phase I, the temperature increased while the pressure was held constant. During phase II, the temperature decreased while the pressure increased.
(C) During phase I, the temperature was held constant while the pressure increased. During phase II, the temperature and pressure both decreased.
(D) During phase I, the temperature and pressure both increased. During phase II, the temperature was held constant while the pressure decreased.

3. A solution of sulfurous acid, H_2SO_3, is present in an aqueous solution. Which of the following represents the concentrations of three different ions in solution?

    (A) $[SO_3^{2-}] > [HSO_3^-] > [H_2SO_3]$
    (B) $[H_2SO_3] > [HSO_3^-] > [SO_3^{2-}]$
    (C) $[H_2SO_3] > [HSO_3^-] = [SO_3^{2-}]$
    (D) $[SO_3^{2-}] = [HSO_3^-] > [H_2SO_3]$

4. $$2\,NO\,(g) + Br_2\,(g) \rightleftharpoons 2\,NOBr\,(g)$$

    The above experiment was performed several times, an the following data is gathered:

Trial	$[NO]_{init}$ (M)	$[Br_2]_{init}$ (M)	Initial Rate of Reaction (M/min)
1	0.20 M	0.10 M	5.20×10^{-3}
2	0.20 M	0.20 M	1.04×10^{-2}
3	0.40 M	0.10 M	2.08×10^{-2}

    What is the rate law for this reaction?

    (A) rate $= k[NO][Br_2]^2$
    (B) rate $= k[NO]^2[Br_2]^2$
    (C) rate $= k[NO][Br_2]$
    (D) rate $= k[NO]^2[Br_2]$

5. $SF_4\,(g) + H_2O\,(l) \rightarrow SO_2\,(g) + 4\,HF\,(g)\quad \Delta H = -828$ kJ/mol

    Which of the following statements accurately describes the above reaction?

    (A) The entropy of the reactants exceeds that of the products.
    (B) $H_2O\,(l)$ will always be the limiting reagent.
    (C) This reaction is never thermodynamically favored.
    (D) The bond strength of the reactants exceeds that of the products.

Use the following information to answer questions 6-9.

20.0 mL of 1.0 M Na_2CO_3 is placed in a beaker and titrated with a solution of 1.0 M $Ca(NO_3)_2$, resulting in the creation of a precipitate. The conductivity of the solution is measured as the titration progresses.

6. How much $Ca(NO_3)_2$ must be added to reach the equivalence point?

    (A) 10.0 mL
    (B) 20.0 mL
    (C) 30.0 mL
    (D) 40.0 mL

GO ON TO THE NEXT PAGE.

7. Which of the following diagrams correctly shows the species present in the solution in significant amounts at the equivalence point?

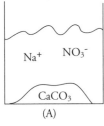

(A)

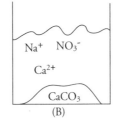

(B)

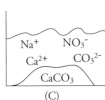

(C)

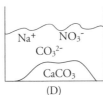

(D)

8. What will happen to the conductivity of the solution after additional $Ca(NO_3)_2$ is added past the equivalence point?

(A) The conductivity will increase as additional ions are being added to the solution.
(B) The conductivity will stay constant as the precipitation reaction has gone to completion.
(C) The conductivity will decrease as the solution will be diluted with the addition of additional $Ca(NO_3)_2$.
(D) The conductivity will stay constant as equilibrium has been established.

9. If the experiment were repeated and the Na_2CO_3 was diluted to 40.0 mL with distilled water prior to the titration, how would that effect the volume of $Ca(NO_3)_2$ needed to reach the equivalence point?

(A) It would be cut in half.
(B) It would decrease by a factor of 1.5.
(C) It would double.
(D) It would not change.

10.
$$2\ CO\ (g) + O_2\ (g) \rightarrow 2\ CO_2\ (g)$$

2.0 mol of CO (g) and 2.0 mol of O_2 (g) are pumped into a rigid, evacuated 4.0-L container, where they react to form CO_2 (g). Which of the following values does NOT represent a potential set of concentrations for each gas at a given point during the reaction?

	CO	O_2	CO_2
(A)	0.5	0.5	0
(B)	0	0.25	0.5
(C)	0.25	0.25	0.5
(D)	0.25	0.38	0.25

11. Neutral atoms of chlorine are bombarded by high-energy photons, causing the ejection of electrons from the various filled subshells. Electrons originally from which subshell would have the highest velocity after being ejected?

(A) $1s$
(B) $2p$
(C) $3p$
(D) $3d$

12. A sample of oxygen gas at 50 °C is heated, reaching a final temperature of 100 °C. Which statement best describes the behavior of the gas molecules?

(A) Their velocity increases by a factor of two.
(B) Their velocity increases by a factor of four.
(C) Their kinetic energy increases by a factor of 2.
(D) Their kinetic energy increases by a factor of less than 2.

13. The average mass, in grams, of one mole of carbon atoms is equal to:

(A) the average mass of a single carbon atom, measured in amus.
(B) the ratio of the number of carbon atoms to the mass of a single carbon atom.
(C) the number of carbon atoms in one amu of carbon.
(D) the mass, in grams, of the most abundant isotope of carbon.

GO ON TO THE NEXT PAGE.

14.

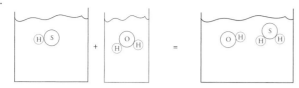

The diagram above best represents which type of reaction?

(A) Acid/base
(B) Oxidation/reduction
(C) Precipitation
(D) Decomposition

15. Which of the following is true for all bases?

(A) All bases donate OH^- ions into solution.
(B) Only strong bases create solutions in which OH^- ions are present.
(C) Only strong bases are good conductors when dissolved in solution.
(D) For weak bases, the concentration of the base exceeds the concentration of OH^- ions in solution.

Use the following information to answer questions 16-18.

$14\ H^+\ (aq) + Cr_2O_7^{2-}\ (aq) + 3\ Ni\ (s) \rightarrow 2\ Cr^{3+}\ (aq) + 3\ Ni^{2+}\ (aq) + 7\ H_2O\ (l)$

In the above reaction, a piece of solid nickel is added to a solution of potassium dichromate.

16. Which species is being oxidized and which is being reduced?

	Oxidized	Reduced
(A)	$Cr_2O_7^{2-}\ (aq)$	$Ni\ (s)$
(B)	$Cr^{3+}\ (aq)$	$Ni^{2+}\ (aq)$
(C)	$Ni\ (s)$	$Cr_2O_7^{2-}\ (aq)$
(D)	$Ni^{2+}\ (aq)$	$Cr^{3+}\ (aq)$

17. How many moles of electrons are transferred when 1 mole of potassium dichromate is mixed with 3 mol of nickel?

(A) 2 moles of electrons
(B) 3 moles of electrons
(C) 5 moles of electrons
(D) 6 moles of electrons

18. How would the pH of the solution change as the reaction progresses?

(A) It increases until the solution becomes basic.
(B) It increases, but the solution remains acidic.
(C) It decreases until the solution becomes basic.
(D) It decreases, but the solution remains acidic.

19. A sample of an unknown chloride compound was dissolved in water, and then titrated with excess $Pb(NO_3)_2$ to create a precipitate. After drying, it is determined there are 0.0050 mol of precipitate present. What mass of chloride is present in the original sample?

(A) 0.177 g
(B) 0.355 g
(C) 0.522 g
(D) 0.710 g

20. Valence electrons in which of the atoms represented below would have the highest amount of potential energy relative to the nucleus?

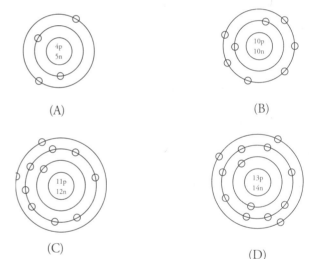

21. The bond length between any two nonmetal atoms is achieved under which of the following distances?

(A) Where the energy of interaction between the atoms is at its minimum value.
(B) Where the nuclei of each atom exhibits the strongest attraction to the electrons of the other atom.
(C) The point at which the attractive and repulsive forces between the two atoms are equal.
(D) The closest point at which a valence electron from one atom can transfer to the other atom.

GO ON TO THE NEXT PAGE.

22. Hydrogen fluoride, HF, is a liquid at 15 °C. All other hydrogen halides (represented by HX, where X is any other halogen) are gases at the same temperature. Why?

    (A) Fluorine has a very high electronegativity; therefore the H-F bond is stronger than any other H-X bond.
    (B) HF is smaller than any other H-X molecule; therefore it exhibits stronger London dispersion forces.
    (C) The dipoles in a HF molecule exhibit a particularly strong attraction force to the dipoles in other HF molecules.
    (D) The H-F bond is the most ionic in character compared to all other hydrogen halides.

23.

	Initial pH	pH after NaOH addition
Acid 1	3.0	3.5
Acid 2	3.0	5.0

Two different acids with identical pH are placed in separate beakers. Identical portions of NaOH are added to each beaker, and the resulting pH is indicated in the table above. What can be determined about the strength of each acid?

    (A) Acid 1 is a strong acid and acid 2 is a weak acid because acid 1 resists change in pH more effectively.
    (B) Acid 1 is a strong acid and acid 2 is a weak acid because the NaOH is more effective at neutralizing the acid 2.
    (C) Acid 1 is a weak acid and acid 2 is a strong acid because the NaOH must accept electrons from a larger percentage of acid 1 in order to neutralize it.
    (D) Acid 1 is a weak acid and acid 2 is a strong acid because the concentration of the hydrogen ions will be greater in acid 2 after the NaOH addition.

24. A stock solution of 12.0 M sulfuric acid is made available. What is the best procedure to make up 100. mL of 4.0 M sulfuric acid using the stock solution and water prior to mixing?

    (A) Add 33.3 mL of water to then flask, then add 66.7 mL of 12.0 M acid.
    (B) Add 33.3 mL of 12.0 M acid to then flask, then dilute it with 66.7 mL of water.
    (C) Add 67.7 mL of 12.0 M acid to the flask, then dilute it with 33.3 mL of water.
    (D) Add 67.7 mL of water to the flask, then add 33.3 mL of 12.0 M acid.

Use the following data to answer questions 25-29

The enthalpy values for the several reactions are as follows:

(I) $C (s) + H_2O (g) \rightarrow CH_4 (g) + H_2 (g)$
$\Delta H = 131$ kJ/mol$_{rxn}$

(II) $CH_4 (g) + H_2O (g) \rightarrow 3 H_2 (g) + CO (g)$
$\Delta H = 206$ kJ/mol$_{rxn}$

(III) $CO (g) + H_2O (g) \rightarrow CO_2 (g) + H_2 (g)$
$\Delta H = -41$ kJ/mol$_{rxn}$

(IV) $CH_4 (g) + 2 O_2 (g) \rightarrow CO_2 (g) + H_2O (l)$
$\Delta H = -890$ kJ/mol$_{rxn}$

25. In which of the the reactions does the energy of the bonds broken exceed the energy of the bonds formed by the greatest amount?

    (A) Reaction I
    (B) Reaction II
    (C) Reaction III
    (D) Reaction IV

26. In which of the reactions is the value for ΔS the most positive?

    (A) Reaction I
    (B) Reaction II
    (C) Reaction III
    (D) Reaction IV

27. Regarding reaction I, how would the addition of a catalyst effect the enthalpy and entropy changes for this reaction?

	Enthalpy	Entropy
(A)	Decrease	Decrease
(B)	Decrease	No Change
(C)	No Change	Decrease
(D)	No Change	No Change

GO ON TO THE NEXT PAGE.

28.

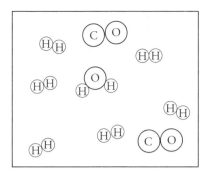

Regarding reaction II, to achieve the products present in the above diagram how many moles of each reactant must be present prior to the reaction?

(A) 1.0 mol of CH_4 and 2.0 mol of H_2O
(B) 2.0 mol of CH_4 and 2.0 mol of H_2O
(C) 2.0 mol of CH_4 and 3.0 mol of H_2O
(D) 3.0 mol of CH_4 and 2.0 mol of H_2O

29. Regarding reaction IV, how much heat is absorbed or released when 2.0 mol of CH_4 (g) reacts with 2.0 mol of O_2 (g) ?

(A) 890 kJ of heat is released.
(B) 890 kJ of heat is absorbed.
(C) 1780 kJ of heat is released.
(D) 1780 kJ of heat is absorbed.

---

30. London dispersion forces are caused by

(A) temporary dipoles created by the position of electrons around the nuclei in a molecule.
(B) the three-dimensional intermolecular bonding present in all covalent substances.
(C) the uneven electron-to-proton ratio found on individual atoms of a molecule.
(D) the electronegativity differences between the different atoms in a molecule.

31. What is the general relationship between temperature and entropy for diatomic substances?

(A) They are completely independent of each other, temperature has no effect on entropy.
(B) There is a direct relationship, as at higher temperatures there is an increase in energy dispersal.
(C) There is an inverse relationship, as at higher temperatures substances are more likely to be in a gaseous state.
(D) It depends on the specific substance and the strength of the intermolecular forces between individual molecules.

32. Which of the following pairs of ions would make the best buffer with a basic pH?

(K_a for $HC_3H_2O_2 = 1.75 \times 10^{-5}$ and K_a for $HPO_4^{2-} = 4.8 \times 10^{-13}$)

(A) H_2SO_4 and Na_2SO_4
(B) HPO_4^{2-} and $H_2PO_4^-$
(C) $HC_3H_2O_2$ and $NaC_3H_2O_2$
(D) LiOH and NaOH

33. A solution contains a mixture of four different compounds: KCl (aq), $Fe(NO_3)_3$ (aq), $MgSO_4$ (aq), and N_2H_4 (aq) Which of these compounds would be easiest to separate via distillation?

(A) KCl (aq)
(B) $Fe(NO_3)_3$ (aq)
(C) $MgSO_4$ (aq)
(D) N_2H_4 (aq)

34.

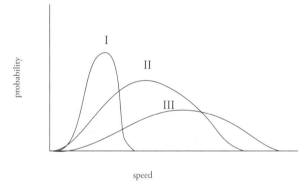

Identify the three gases represented on the Maxwell-Boltzmann diagram above. Assume all gases are at the same temperature.

	I	II	III
(A)	H_2	N_2	F_2
(B)	H_2	F_2	N_2
(C)	F_2	N_2	H_2
(D)	N_2	F_2	H_2

GO ON TO THE NEXT PAGE.

35. A sample of solid $MgCl_2$ would be most soluble in which of the following solutions?

(A) $LiOH$ (aq)
(B) CBr_4 (aq)
(C) $Mg(NO_3)_2$ (aq)
(D) $AlCl_3$ (aq)

36. Most transition metals share a common oxidation state of +2. Which of the following best explains why?

(A) Transition metals all have a minimum of two unpaired electrons.
(B) Transition metals have unstable configurations and are very reactive.
(C) Transition metals tend to gain electrons when reacting with other elements.
(D) Transition metals will lose their outermost s-block electrons when forming bonds.

37. $2 Ag^+$ (aq) + Fe (s) → $2 Ag$ (s) + Fe^{2+} (aq)

Which of the following would cause an increase in potential in the voltaic cell described by the above reaction?

(A) increasing Fe^{2+}
(B) adding more Fe (s)
(C) decreasing Fe^{2+}
(D) removing some Fe (s)

Use the following information to answer questions 38-41.

Consider the Lewis structures for the following molecules:

$$CO_2, CO_3^{2-}. NO_2^-, \text{ and } NO_3^-$$

38. Which molecule would have the shortest bonds?

(A) CO_2
(B) CO_3^{2-}
(C) NO_2^-
(D) NO_3^-

39. Which molecules are best represented by multiple resonance structures?

(A) CO_2 and CO_3^{2-}
(B) NO_2^- and NO_3^-
(C) CO_3^{2-} and NO_3^-
(D) CO_3^{2-}, NO_2^- and NO_3^-

40. Which molecule or molecules exhibit sp^2 hybridization around the central carbon atom?

(A) CO_2 and CO_3^{2-}
(B) NO_2^- and NO_3^-
(C) CO_3^{2-} and NO_3^-
(D) CO_3^{2-}, NO_2^- and NO_3^-

41. Which molecule would have the smallest bond angle between terminal atoms?

(A) CO_2
(B) CO_3^{2-}
(C) NO_2^-
(D) NO_3^-

42. NH_4^+ (aq) + NO_2^- (aq) → N_2 (g) + $2 H_2O$ (l)

Increasing the temperature of the above reaction will increase the rate of reaction.

Which of the following is NOT a reason that increased temperature increases reaction rate?

(A) The reactants will be more likely to overcome the activation energy.
(B) The number of collisions between reactant molecules will increase.
(C) A higher distribution of reactant molecules will have high velocities.
(D) Alternate reaction pathways become available at higher temperatures.

43. Which of the following diagrams best represents what is happening on a molecular level when NaCl dissolves in water?

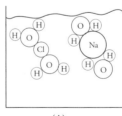

(A)

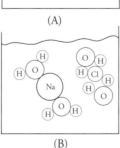

(B)

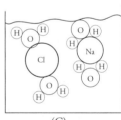

(C)

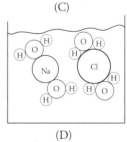

(D)

GO ON TO THE NEXT PAGE.

44. Nitrous acid, HNO_2, has a pK_a value of 3.3. If a solution of nitric acid is found to have a pH of 4.2, what can be said about the concentration of the conjugate acid/base pair found in solution?

    (A) $[HNO_2] > [NO_2^-]$
    (B) $[NO_2^-] > [HNO_2]$
    (C) $[H_2NO_2^+] > [HNO_2]$
    (D) $[HNO_2] > [H_2NO_2^+]$

45. Which of the following processes is an irreversible reaction?

    (A) $CH_4 (g) + O_2 (g) \rightarrow CO_2 (g) + H_2O (l)$
    (B) $HCN (aq) + H_2O (l) \rightarrow CN^- (aq) + H_3O^+ (aq)$
    (C) $Al(NO_3)_3 (aq) \rightarrow Al^{3+} (aq) + 3\ NO_3^- (aq)$
    (D) $2\ Ag^+ (aq) + Ti (s) \rightarrow 2\ Ag (s) + Ti^{2+} (aq)$

Use the following information to answer questions 46-50.

A sample of H_2S gas is placed in an evacuated, sealed container and heated until the following decomposition reaction occurs at 1000 K:

$2\ H_2S (g) \rightarrow 2\ H_2 (g) + S_2 (g)$ $K_c = 1.0 \times 10^{-6}$

46. Which of the following represents the equilibrium constant for this reaction?

    (A) $K_c = \dfrac{[H_2]^2 [S_2]}{[H_2S]^2}$

    (B) $K_c = \dfrac{[H_2S]^2}{[H_2]^2 [S_2]}$

    (C) $K_c = \dfrac{2[H_2][S_2]}{2[H_2S]}$

    (D) $K_c = \dfrac{2[H_2S]}{2[H_2][S_2]}$

47. Which of the following graphs would best represent the change in concentration of the various species involved in the reaction over time?

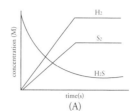
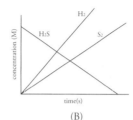

(A)

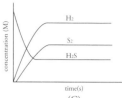
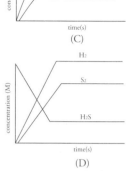

(C)

(B)

(D)

48. Which option best describes what will happen to the reaction rates if the pressure on the system is increased after it has reached equilibrium?

    (A) The rate of both the forward and reverse reactions will increase.
    (B) The rate of the forward reaction will increase while the rate of the reverse reaction decreases.
    (C) The rate of the forward reaction will decrease while the rate of the reverse reaction increases.
    (D) Neither the rate of the forward nor reverse reactions will change.

49. If, at a given point in the reaction, the value for the reaction quotient Q is determined to be 2.5×10^{-8}, which of the following is occurring?

    (A) The concentration of the reactant is decreasing while the concentration of the products is increasing.
    (B) The concentration of the reactant is increasing while the concentration of the products is decreasing.
    (C) The system has passed the equilibrium point and the concentration of all species involved in the reaction will remain constant.
    (D) The concentrations of all species involved are changing at the same rate.

GO ON TO THE NEXT PAGE.

50. As the reaction progresses at a constant temperature of 1000K, how does the value for the Gibbs free energy constant for the reaction change?

    (A) It stays constant.
    (B) It increases exponentially.
    (C) It increases linearly.
    (D) It decreases exponentially.

51. An unknown substance is found to have a high melting point. In addition, it is a poor conductor of electricity and does not dissolve in water. The substance most likely contains

    (A) ionic bonding.
    (B) nonpolar covalent bonding.
    (C) covalent network bonding.
    (D) metallic bonding.

52. Which of the following best explains why the ionization of atoms can occur during photoelectron spectroscopy, even though ionization is not a thermodynamically favored process?

    (A) It is an exothermic process due to the release of energy as an electron is liberated from the Coulombic attraction holding it to the nucleus.
    (B) The entropy of the system increases due to the separation of the electron from its atom.
    (C) Energy contained in the light can be used to overcome the Coulombic attraction between electrons and the nucleus.
    (D) The products of the ionization are at a lower energy state than the reactants.

53.
$$2H_2O_2\ (aq) \longrightarrow 2H_2O\ (l) + O_2\ (g)$$

H–O–O–H (x8)	H–O–O–H (x4)	H–O–O–H (x2)
	H–O–H (x4)	H–O–H (x6)
	O=O (x2)	O=O (x3)
$t = 0s$	$t = 200s$	$t = 400s$

The above diagrams shows the decomposition of hydrogen peroxide in a sealed container in the presence of a catalyst. What is the overall order for the reaction?

    (A) Zero order
    (B) First order
    (C) Second order
    (D) Third order

54.

One of the resonance structures for the nitrite ion is shown above. What is the formal charge on each atom?

	O_x	N	O_y
(A)	−1	+1	−1
(B)	+1	−1	0
(C)	0	0	−1
(D)	−1	0	0

Use the following information to answer questions 55-57.

Atoms of four elements are examined: carbon, nitrogen, neon, and sulfur.

55. Atoms of which element would have the strongest magnetic moment?

    (A) carbon
    (B) nitrogen
    (C) neon
    (D) sulfur

56. Atoms of which element are most likely to form a structure with the formula XF_6 (where X is one of the four atoms)

    (A) carbon
    (B) nitrogen
    (C) neon
    (D) sulfur

57. Which element would have a photoelectron spectra in which the peak representing electrons with the lowest ionization energy would be a three times higher than all other peaks?

    (A) carbon
    (B) nitrogen
    (C) neon
    (D) sulfur

GO ON TO THE NEXT PAGE.

58. The diagram below supports which of the following conclusions about the reaction shown below?

$$\ddot{N} \equiv \ddot{N} \quad + \quad \begin{array}{c} H\text{-}H \\ H\text{-}H \\ H\text{-}H \end{array} \quad \longrightarrow \quad H{\diagup}\overset{\overset{\displaystyle \cdot\cdot}{N}}{\underset{H}{|}}{\diagdown}H$$

(A) There is an increase in entropy.
(B) Mass is conserved in all chemical reactions.
(C) The bond length increases after the reaction.
(D) The enthalpy value is positive.

59.
$$NO_2 + O_3 \rightarrow NO_3 + O_2 \qquad \text{slow}$$
$$NO_3 + NO_2 \rightarrow N_2O_5 \qquad \text{fast}$$

A proposed reaction mechanism for the reaction of nitrogen dioxide and ozone is detailed above. Which of the following is the rate law for the reaction?

(A) rate $= k[NO_2][O_3]$
(B) rate $= k[NO_3][NO_2]$
(C) rate $= k[NO_2]^2[O_3]$
(D) rate $= k[NO_3][O_2]$

60.

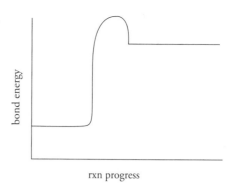

rxn progress

The concentrations of the reactants and products in the reaction represented by the above coordinate are found are found to be changing very slowly. Which of the following statements best describes the reaction a given that the reaction is exergonic? ($\Delta G < 0$)

(A) The reaction is under kinetic control.
(B) The reaction has reached a state of equilibrium.
(C) The reaction is highly exothermic in nature.
(D) The addition of heat will increase the rate of reaction significantly.

END OF SECTION 1

CHEMISTRY
SECTION II
Time—1 hour and 30 minutes

General Instructions

Calculators, including those with programming and graphing capabilities, may be used in Part A. However, calculators with typewriter-style (QWERTY) keyboards are NOT permitted.

Pages containing a periodic table, the electrochemical series, and equations commonly used in chemistry will be available for your use.

You may write your answers with either a pen or a pencil. Be sure to write CLEARLY and LEGIBLY. If you make an error, you may save time by crossing it out rather than trying to erase it.

Write all your answers in the essay booklet. Number your answers as the questions are numbered in the examination booklet.

GO ON TO THE NEXT PAGE.

ADVANCED PLACEMENT CHEMISTRY EQUATIONS AND CONSTANTS

ATOMIC STRUCTURE

$$E = h\nu \qquad\qquad c = \lambda\nu$$

$$\lambda = \frac{h}{mv} \qquad\qquad p = mv$$

$$E_n = \frac{-2.178 \times 10^{-18}}{n^2} \text{ joule}$$

EQUILIBRIUM

$$K_a = \frac{[H^-][A^-]}{[HA]}$$

$$K_b = \frac{[OH^-][HB^+]}{[B]}$$

$$K_w = [OH^-][H^+] = 10^{-14} @ 25°C$$
$$= K_a \times K_b$$

$$pH = -\log[H^+], pOH = -\log[OH^-]$$

$$14 = pH + pOH$$

$$pH = pK_a + \log\frac{[A^-]}{[HA]}$$

$$pOH = pK_b + \log\frac{[HB^+]}{[B]}$$

$$pK_a = -\log K_a, pK_b = -\log K_b$$

$$K_p = K_c(RT)^{\Delta n},$$

where Δn = moles product gas – moles reactant gas

THERMOCHEMISTRY / KINETICS

$$\Delta S° = \sum S°_{products} - \sum S°_{reactants}$$

$$\Delta H° = \sum \Delta H°_{f\ products} - \sum \Delta H°_{f\ reactants}$$

$$\Delta G° = \sum \Delta G°_{f\ products} - \sum \Delta G°_{f\ reactants}$$

$$\Delta G° = \Delta H° - T\Delta S°$$
$$= -RT \ln K = -2.303\ RT \log K$$
$$= -nFE°$$

$$\Delta G = \Delta G° + RT \ln Q = \Delta G° + 2.303\ RT \log Q$$
$$q = mc\Delta T$$
$$C_p = \frac{\Delta H}{\Delta T}$$

$$\ln[A]_t = -kt + \ln[A]_o$$

$$\frac{1}{[A]_t} = kt + \frac{1}{[A]_o}$$

$$\ln k = \frac{-E_a}{R}\left(\frac{1}{T}\right) + \ln A$$

E = energy $\qquad v$ = velocity

ν = frequency $\qquad n$ = principal quantum number

λ = wavelength $\qquad m$ = mass

p = momentum

Speed of light, $c = 3.00 \times 10^8$ m s^{-1}

Planck's constant, $h = 6.63 \times 10^{-34}$ J s

Boltzmann's constant, $k = 1.38 \times 10^{-23}$ J K^{-1}

Avogadro's number $= 6.022 \times 10^{23}$ mol^{-1}

Electron charge, $e = -1.602 \times 10^{-19}$ coulomb

1 electron volt/atom $= 96.5$ kJ mol^{-1}

Equilibrium Constants

K_a (weak acid)
K_b (weak base)
K_w (water)
K_p (gas pressure)
K_c (molar concentrations)

$S°$ = standard entropy
$H°$ = standard enthalpy
$G°$ = standard free energy
$E°$ = standard reduction potential
T = temperature
n = moles
m = mass
q = heat
c = specific heat capacity
C_p = molar heat capacity at constant pressure
E_a = activation energy
k = rate constant
A = frequency factor

Faraday's Constant, F = 96,500 coulombs per mole
of electrons

Gas Constant, R = 8.31 J mol^{-1} K^{-1}
= 0.0821 L atm mol^{-1} K^{-1}
= 8.31 volt coulomb mol^{-1} K^{-1}

GO ON TO THE NEXT PAGE.

MATERIAL IN THE FOLLOWING TABLE AND IN THE TABLES ON THE NEXT 3 PAGES MAY BE USEFUL IN ANSWERING THE QUESTIONS IN THIS SECTION OF THE EXAMINATION.

PERIODIC CHART OF THE ELEMENTS

1 H 1.0																	2 He 4.0
3 Li 6.9	4 Be 9.0											5 B 10.8	6 C 12.0	7 N 14.0	8 O 16.0	9 F 19.0	10 Ne 20.2
11 Na 23.0	12 Mg 24.3											13 Al 27.0	14 Si 28.1	15 P 31.0	16 S 32.1	17 Cl 35.5	18 Ar 39.9
19 K 39.1	20 Ca 40.1	21 Sc 45.0	22 Ti 47.9	23 V 50.9	24 Cr 52.0	25 Mn 54.9	26 Fe 55.8	27 Co 58.9	28 Ni 58.7	29 Cu 63.5	30 Zn 65.4	31 Ga 69.7	32 Ge 72.6	33 As 74.9	34 Se 79.0	35 Br 79.9	36 Kr 83.8
37 Rb 85.5	38 Sr 87.6	39 Y 88.9	40 Zr 91.2	41 Nb 92.9	42 Mo 95.9	43 Tc (98)	44 Ru 101.1	45 Rh 102.9	46 Pd 106.4	47 Ag 107.9	48 Cd 112.4	49 In 114.8	50 Sn 118.7	51 Sb 121.8	52 Te 127.6	53 I 126.9	54 Xe 131.3
55 Cs 132.9	56 Ba 137.3	57 *La 138.9	72 Hf 178.5	73 Ta 180.9	74 W 183.9	75 Re 186.2	76 Os 190.2	77 Ir 192.2	78 Pt 195.1	79 Au 197.0	80 Hg 200.6	81 Tl 204.4	82 Pb 207.2	83 Bi 209.0	84 Po (209)	85 At (210)	86 Rn (222)
87 Fr (223)	88 Ra 226.0	89 †Ac 227.0															

*Lanthanum Series

58 Ce 140.1	59 Pr 140.9	60 Nd 144.2	61 Pm (145)	62 Sm 150.4	63 Eu 152.0	64 Gd 157.3	65 Tb 158.9	66 Dy 162.5	67 Ho 164.9	68 Er 167.3	69 Tm 168.9	70 Yb 173.0	71 Lu 175.0

†Actinium Series

90 Th 232.0	91 Pa 231.0	92 U 238.0	93 Np 237.0	94 Pu (244)	95 Am (243)	96 Cm (247)	97 Bk (247)	98 Cf (251)	99 Es (252)	100 Fm (258)	101 Md (258)	102 No (259)	103 Lr (260)

DO NOT DETACH FROM BOOK.

GO ON TO THE NEXT PAGE.

ADVANCED PLACEMENT CHEMISTRY EQUATIONS AND CONSTANTS

GASES, LIQUIDS, AND SOLUTIONS

$$PV = nRT$$

$$\left(P + \frac{n^2 a}{V^2}\right)(V - nb) = nRT$$

$$P_A = P_{total} \cdot X_A, \text{ where } X_A = \frac{\text{moles A}}{\text{total moles}}$$

$$P_{total} = P_A + P_B + P_C{}^+ \ldots$$

$$n = \frac{m}{M}$$

$$K = {}^\circ C + 273$$

$$\frac{P_1 V_1}{T_1} = \frac{P_2 V_2}{T_2}$$

$$D = \frac{m}{V}$$

$$U_{rms} = \sqrt{\frac{3kT}{m}} = \sqrt{\frac{3RT}{M}}$$

$$KE \text{ per molecule} = \frac{1}{2} mv^2$$

$$KE \text{ per mole} = \frac{3}{2} RT$$

$$\frac{r_1}{r_2} = \sqrt{\frac{M_2}{M_1}}$$

molarity, M = moles solute per liter solution

molality = moles solute per kilogram solvent

$$\Delta T_f = iK_f \times \text{molality}$$

$$\Delta T_b = iK_b \times \text{molality}$$

$$\pi = MRT$$

$$A = abc$$

P = pressure
V = volume
T = temperature
n = number of moles
D = density
m = mass
v = velocity

U_{rms} = root-mean-square speed
KE = kinetic energy
r = rate of effusion
M = molar mass
π = osmotic pressure
i = van't Hoff factor
K_f = molal freezing-point depression constant
K_b = molal boiling-point elevation constant
A = absorbance
a = molar absorptivity
b = path length
c = concentration
Q = reaction quotient
l = current (amperes)
q = charge (coulombs)
t = time (seconds)
E° = standard reduction potential
K = equilibrium constant

OXIDATION REDUCTION; ELECTROCHEMISTRY

$$Q = \frac{[C]^c [D]^d}{[A]^a [B]^b} \text{ where } a \text{ A} + b \text{ B} \rightarrow c \text{ C} + d \text{ D}$$

$$I = \frac{q}{t}$$

$$E_{cell} = E^\circ{}_{cell} - \frac{RT}{nF} \ln Q = E^\circ{}_{cell} - \frac{0.0592}{n} \log Q \ @ \ 25^\circ C$$

$$\log K = \frac{nE^\circ}{0.0592}$$

Gas constant, $R = 8.31 \text{ J mol}^{-1} \text{K}^{-1}$
$= 0.0821 \text{ L atm mol}^{-1} \text{ K}^{-1}$
$= 8.31 \text{ volt coulomb mol}^{-1} \text{K}^{-1}$
Boltzmann's constant, $k = 1.38 \times 10^{-23} \text{ J K}^{-1}$
K_f for $H_2O = 1.86 \text{ K kg mol}^{-1}$
K_b for $H_2O = 0.512 \text{ K kg mol}^{-1}$
$1 \text{ atm} = 760 \text{ mm Hg}$
$= 760 \text{ torr}$
STP $= 0.000^\circ C$ and 1.000 atm
1 faraday, $F = 96{,}500$ coulombs per mole
of electrons

GO ON TO THE NEXT PAGE.

STANDARD REDUCTION POTENTIALS IN AQUEOUS SOLUTION AT 25 C (in V)

$F_2(g) + 2 e^-$	$\rightarrow$	$2 F^-$	2.87
$Co^{3+} + e^-$	$\rightarrow$	Co^{2+}	1.82
$Au^{3+} + 3e^-$	$\rightarrow$	$Au(s)$	1.50
$Cl_2(g) + 2 e^-$	$\rightarrow$	$2 Cl^-$	1.36
$O_2(g) + 4H^+ + 4 e^-$	$\rightarrow$	$2 H_2O$	1.23
$Br_2(l) + 2 e^-$	$\rightarrow$	$2 Br^-$	1.07
$2 Hg^{2+} + 2 e^-$	$\rightarrow$	Hg_2^{2+}	0.92
$Hg^{2+} + 2 e^-$	$\rightarrow$	$Hg(l)$	0.85
$Ag^+ + e^-$	$\rightarrow$	$Ag(s)$	0.80
$Hg_2^{2+} + 2 e^-$	$\rightarrow$	$2 Hg(l)$	0.79
$Fe^{3+} + e^-$	$\rightarrow$	Fe^{2+}	0.77
$I_2(s) + 2 e^-$	$\rightarrow$	$2 I^-$	0.53
$Cu^+ + e^-$	$\rightarrow$	$Cu(s)$	0.52
$Cu^{2+} + 2 e^-$	$\rightarrow$	$Cu(s)$	0.34
$Cu^{2+} + e^-$	$\rightarrow$	Cu^+	0.15
$Sn^{4+} + 2 e^-$	$\rightarrow$	Sn^{2+}	0.15
$S(s) + 2 H^+ + 2 e^-$	$\rightarrow$	H_2S	0.14
$2 H^+ + 2 e^-$	$\rightarrow$	$H_2(g)$	0.00
$Pb^{2+} + 2 e^-$	$\rightarrow$	$Pb(s)$	−0.13
$Sn^{2+} + 2 e^-$	$\rightarrow$	$Sn(s)$	−0.14
$Ni^{2+} + 2 e^-$	$\rightarrow$	$Ni(s)$	−0.25
$Co^{2+} + 2 e^-$	$\rightarrow$	$Co(s)$	−0.28
$Tl^+ + e^-$	$\rightarrow$	$Tl(s)$	−0.34
$Cd^{2+} + 2 e^-$	$\rightarrow$	$Cd(s)$	−0.40
$Cr^{3+} + e^-$	$\rightarrow$	Cr^{2+}	−0.41
$Fe^2 + 2 e^-$	$\rightarrow$	$Fe(s)$	−0.44
$Cr^{3+} + 3 e^-$	$\rightarrow$	$Cr(s)$	−0.74
$Zn^{2+} + 2 e^-$	$\rightarrow$	$Zn(s)$	−0.76
$Mn^{2+} + 2 e^-$	$\rightarrow$	$Mn(s)$	−1.18
$Al^{3+} + 3 e^-$	$\rightarrow$	$Al(s)$	−1.66
$Be^{2+} + 2 e^-$	$\rightarrow$	$Be(s)$	−1.70
$Mg^{2+} + 2 e^-$	$\rightarrow$	$Mg(s)$	−2.37
$Na^+ + e^-$	$\rightarrow$	$Na(s)$	−2.71
$Ca^{2+} + 2 e^-$	$\rightarrow$	$Ca(s)$	−2.87
$Sr^{2+} + 2 e^-$	$\rightarrow$	$Sr(s)$	−2.89
$Ba^{2+} + 2 e^-$	$\rightarrow$	$Ba(s)$	−2.90
$Rb^+ + e^-$	$\rightarrow$	$Rb(s)$	−2.92
$K^+ + e^-$	$\rightarrow$	$K(s)$	−2.92
$Cs^+ + e^-$	$\rightarrow$	$Cs(s)$	−2.92
$Li^+ + e^-$	$\rightarrow$	$Li(s)$	−3.05

GO ON TO THE NEXT PAGE.

CHEMISTRY
Section II
(Total time—90 minutes)

Part A
YOU MAY USE YOUR CALCULATOR FOR PART A.

THE METHODS USED AND THE STEPS INVOLVED IN ARRIVING AT YOUR ANSWERS MUST BE SHOWN CLEARLY. It is to your advantage to do this since you may obtain partial credit if you do, and you will receive little or no credit if you do not. Attention should be paid to significant figures.

1. A stock solution of 0.100 M cobalt (II) chloride is used to create several solutions, indicated in the data table below:

Sample	Volume $CoCl_2$ (mL)	Volume H_2O (mL)
1	20.00	0
2	15.00	5.00
3	10.00	10.00
4	5.00	15.00

(a) In order to achieve the degree of accuracy shown in the table above select which of the following pieces of laboratory equipment could be used when measuring out the $CoCl_2$:

150-mL beaker   400-mL beaker   250-mL Earlenmeyer flask

50-mL buret   50-mL graduated cylinder   100-mL graduated cylinder

(b) Calculate the concentration of the $CoCl_2$ in each sample. The solutions are then placed in cuvettes before being inserted into a spectrophotometer calibrated to 560 nm and their values are measured, yielding the data below:

Sample	Absorbance
1	0.485
2	0.364
3	0.243
4	0.121

(c) If gloves are not worn when handling the cuvettes, how might this affect the absorbance values gathered?

(d) If the path length of the cuvette is 1.00 cm, what is the molar absorptivity value for $CoCl_2$ at 560 nm?

(e) On the axes below, plot a graph of absorbance vs. concentration. The *y*-axes scale is set, and be sure to scale the x-axes appropriately

GO ON TO THE NEXT PAGE.

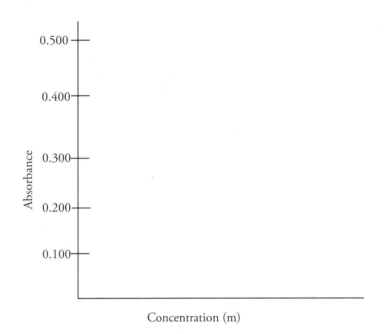

(f)   What would the absorbance values be for $CoCl_2$ solutions at the following concentrations?
   (i)   0.067 M
   (ii) 0.180 M

GO ON TO THE NEXT PAGE.

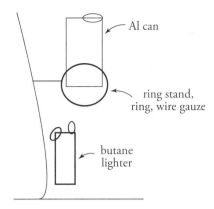

2. A sample of liquid butane (C_4H_{10}) in a pressurized lighter is set up directly beneath an aluminum can, as show in the diagram above. The can contains 100.0 mL of water, and when the butane is ignited the temperature of the water inside the can increases from 25.0 °C to 82.3 °C. The total mass of butane ignited is found to be 0.51 g, the specific heat of water is 4.18 J/g °C, and the density of water is 1.00 g/mL.

   (a) Write the balanced chemical equation for the combustion of butane in air.
   (b) (i) How much heat did the water gain?
        (ii) What is the experimentally determined heat of combustion for butane based on this experiment? Your answer should be in kJ/mol.
   (c) Given butane's density of 0.573 g/mL at 25 °C, calculate how much heat would be emitted if 5.00 mL of it were combusted at that temperature.
   (d) The overall combustion of butane is an exothermic reaction. Explain why this is in terms of bond energies.
   (e) One of the major sources of error in this experiment comes from the heat which is absorbed by the air. Why, then, might it not be a good idea to perform this experiment inside a sealed container to prevent the heat from leaving the system?

GO ON TO THE NEXT PAGE.

3. $$2 N_2O_5 (g) \rightarrow 4 NO_2 (g) + O_2 (g)$$

The data below was gathered for the decomposition of N_2O_5 at 310 K via the equation above.

Time (s)	$[N_2O_5]$ (M)
0	0.250
500.	0.190
1000.	0.145
2000.	0.085

(a)   How does the rate of appearance of NO_2 compare to the rate of disappearance of N_2O_5? Justify your answer.

(b)   The reaction is determined to be first order overall. On the axes below, create a graph of some function of concentration vs. time that will produce a straight line. Label and scale your axes appropriately.

(c)   (i)   What is the rate constant for this reaction? Include units!
       (ii)   What would the concentration of N_2O_5 be at $t = 1500$ s?
       (iii)   What is the half-life of N_2O_5?

(d)   Would the addition of a catalyst increase, decrease, or have no effect on the following variables? Justify your answers.:
       (i)   Rate of disappearance of N_2O_5
       (ii)   Magnitude of the rate constant
       (iii)   Half-life of N_2O_5

GO ON TO THE NEXT PAGE.

Part B

Time—40 minutes

CALCULATORS MAY BE USED FOR PART B

4. Consider the Lewis structures for the following four molecules:

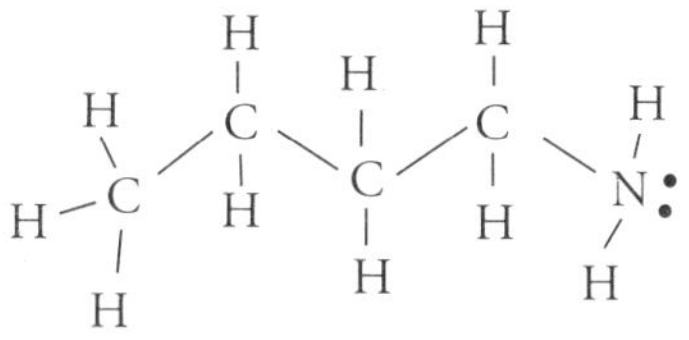

n-Butylamine

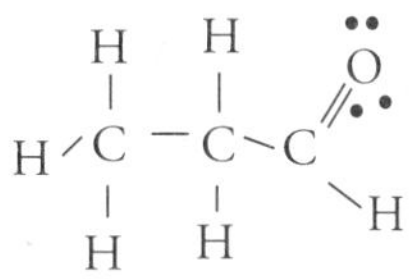

Propanal

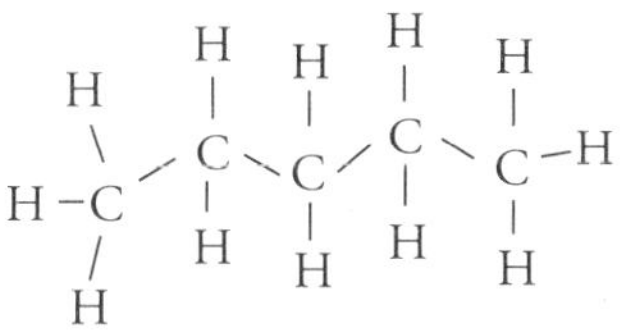

Pentane

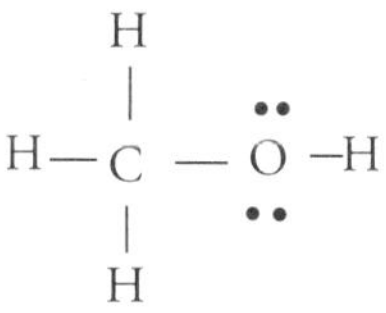

Methanol

(a) All of the substances are liquids at room temperature. Organize them from high to low in terms of boiling points, clearly differentiating between the intermolecular forces in each substance.

(b) On the methanol diagram reproduced below, draw the locations of all dipoles.

(c) n-Butylamine is found to have the lowest vapor pressure at room temperature out of the four liquids. Justify this observation in terms of intermolecular forces.

GO ON TO THE NEXT PAGE.

5. Current is run through a solution of aqueous solution of 1.0 M nickel (II) chloride, and a gas is evolved at the right-hand electrode, as indicated by the diagram below:

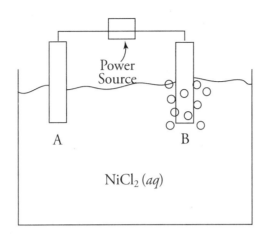

The standard reduction potential for several reactions is given in the following table:

Half-cell	E°_{red}
$Cl_2 (g) + 2e^- \rightarrow 2\ Cl^-$	+1.36 V
$O_2 (g) + 4\ H^+ + 4e^- \rightarrow 2\ H_2O\ (l)$	+1.23 V
$Ni^{2+} + 2e^- \rightarrow Ni\ (s)$	–0.25 V
$2\ H_2O\ (l) + 2e^- \rightarrow H_2 (g) + 2\ OH^-$	–0.83 V

(a) Determine which half-reaction is occurring at each electrode:
  (i) Oxidation:
  (ii) Reduction:
(b) (i) Calculate the standard cell potential for the cell.
  (ii) Calculate the Gibbs free energy value for the cell.
(c) Which electrode in the diagram (A or B) is the cathode, and which is the anode? Justify your answers.

GO ON TO THE NEXT PAGE.

6. Aniline, $C_6H_5NH_2$ is a weak base with $K_b = 3.8 \times 10^{-10}$.

   (a) Write out the reaction which occurs when aniline reacts with water.
   (b) (i) What is the concentration of each species at equilibrium in a solution of 0.25 M $C_6H_5NH_2$?
        (ii) What is the pH value for the solution in b(i)?

7. A rigid, sealed 12.00 L container is filled with 10.00 g each of three different gases: CO_2, NO, and NH_3. The temperature of the gases is held constant 35.0 °C. Assume ideal behavior for all gases.

   (a) (i) What is the mole fraction of each gas?
        (ii) What is the partial pressure of each gas?
   (b) Out of the three gases, molecules of which gas will have the highest velocity? Why?
   (c) Name one circumstance in which the gases might deviate from ideal behavior, and clearly explain the reason for the deviation.

STOP

END OF EXAM

Chapter 13
Practice Test 2
Answers and
Explanations

ANSWER KEY EXAM 2

1.	B	31.	B
2.	B	32.	B
3.	B	33.	D
4.	D	34.	C
5.	D	35.	A
6.	B	36.	D
7.	A	37.	C
8.	A	38.	A
9.	D	39.	D
10.	C	40.	D
11.	C	41.	C
12.	D	42.	D
13.	A	43.	D
14.	A	44.	B
15.	C	45.	A
16.	C	46.	A
17.	D	47.	C
18.	B	48.	C
19.	B	49.	A
20.	B	50.	A
21.	A	51.	C
22.	C	52.	C
23.	C	53.	B
24.	D	54.	D
25.	B	55.	B
26.	A	56.	D
27.	D	57.	C
28.	C	58.	B
29.	A	59.	A
30.	A	60.	A

Section I—Multiple Choice

1. A compound is made up of entirely silicon and oxygen atoms. If there is 14.00 g of silicon and 32.0 g of oxygen present, what is the empirical formula of the compound?

    (A) SiO_2
    (B) SiO_4
    (C) Si_2O
    (D) Si_2O_3

2.

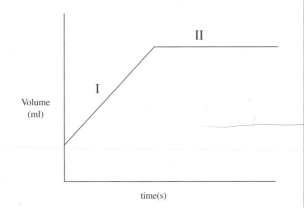

time(s)

The volume of a gas is charted over time, giving the above results. Which of the following options provides a possible explanation of what was happening to the gas during each phase of the graph?

    (A) During phase I, the temperature decreased while the pressure increased. During phase II, the temperature was held constant as the pressure decreased.
    (B) During phase I, the temperature increased while the pressure was held constant. During phase II, the temperature decreased while the pressure increased.
    (C) During phase I, the temperature was held constant while the pressure increased. During phase II, the temperature and pressure both decreased.
    (D) During phase I, the temperature and pressure both increased. During phase II, the temperature was held constant while the pressure decreased.

1. **B** 14.0 g of Si is 0.50 mol, and 32.0 g of oxygen is 2.0 mol. Converting that to a whole number ratio gives us 1 mol of Si for every 4 moles of O.

2. **B** For the volume to increase in Phase I, either the pressure has to decrease or the temperature has to increase (or both). For the volume to remain constant in Phase II, either both pressure and volume have to remain constant, or one has to increase while the other decreases.

3. A solution of sulfurous acid, H_2SO_3, is present in an aqueous solution. Which of the following represents the concentrations of three different ions in solution?

   (A) $[SO_3^{2-}] > [HSO_3^-] > [H_2SO_3]$
   (B) $[H_2SO_3] > [HSO_3^-] > [SO_3^{2-}]$
   (C) $[H_2SO_3] > [HSO_3^-] = [SO_3^{2-}]$
   (D) $[SO_3^{2-}] = [HSO_3^-] > [H_2SO_3]$

3. **B** Weak acids have a very low dissociation value, and for polyprotic weak acids the second dissociation is always weaker than the first.

4.
$$2 NO\ (g) + Br_2\ (g)\ 2\ NOBr\ (g)$$

The above experiment was performed several times, an the following data is gathered:

Trial	$[NO]_{init}$ (M)	$[Br_2]_{init}$ (M)	Initial Rate of Reaction (M/min)
1	0.20 M	0.10 M	5.20×10^{-3}
2	0.20 M	0.20 M	1.04×10^{-2}
3	0.40 M	0.10 M	2.08×10^{-2}

What is the rate law for this reaction?

   (A) rate = $k[NO][Br_2]^2$
   (B) rate = $k[NO]^2[Br_2]^2$
   (C) rate = $k[NO][Br_2]$
   (D) rate = $k[NO]^2[Br_2]$

4. **D** Between trial 1 and 3, the concentration of NO doubled while the Br_2 held constant, and the rate went up by a factor of four. So, the reaction is second order with respect to NO. Between trial 1 and 2, the concentration of NO was held constant while the concentration of Br_2 doubled, and the rate went up by a factor of two. Thus, the reaction is first order with respect to Br_2.

5. $SF_4\ (g) + H_2O\ (l) \rightarrow SO_2\ (g) + 4\ HF\ (g)$ $\Delta H = -828$ kJ/mol

Which of the following statements accurately describes the above reaction?

   (A) The entropy of the reactants exceeds that of the products.
   (B) $H_2O\ (l)$ will always be the limiting reagent.
   (C) This reaction is never thermodynamically favored.
   (D) The bond strength of the reactants exceeds that of the products.

5. **D** One way to calculate enthalpy is by subtracting the strength of the bonds broken in the reactants by the strength of the bonds formed in the products. If the first value is greater than the second value, the overall enthalpy change will be negative.

QUESTIONS	EXPLANATIONS

Use the following information to answer questions 6-9.

20.0 mL of 1.0 M Na_2CO_3 is placed in a beaker and titrated with a solution of 1.0 M $Ca(NO_3)_2$, resulting in the creation of a precipitate. The conductivity of the solution is measured as the titration progresses.

6. How much $Ca(NO_3)_2$ must be added to reach the equivalence point?

   (A) 10.0 mL
   (B) 20.0 mL
   (C) 30.0 mL
   (D) 40.0 mL

7. Which of the following diagrams correctly shows the species present in the solution in significant amounts at the equivalence point?

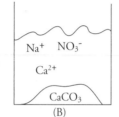

6. **B** Equivalence is reached when the conductivity of the solution reaches it's minimum value, meaning as many of the precipitate ions are out of solution as possible. As the formula of the precipitate is $CaCO_3$, for every mole of sodium carbonate in the beaker, one mole of calcium nitrate will be needed. As both solutions have equal molarity, equal volumes of each would be necessary to ensure the moles are equal.

7. **A** At the equivalence point, the moles of Ca^{2+} and CO_3^{2-} in solution will be negligible.

8. What will happen to the conductivity of the solution after additional $Ca(NO_3)_2$ is added past the equivalence point?

    (A) The conductivity will increase as additional ions are being added to the solution.
    (B) The conductivity will stay constant as the precipitation reaction has gone to completion.
    (C) The conductivity will decrease as the solution will be diluted with the addition of additional $Ca(NO_3)_2$.
    (D) The conductivity will stay constant as equilibrium has been established.

8. **A** The Ca^{2+} ions will have nothing to react with, so they will just remain in solution (as will the extra NO_3^- ions being added), increasing the conductivity.

9. If the experiment were repeated and the Na_2CO_3 was diluted to 40.0 mL with distilled water prior to the titration, how would that effect the volume of $Ca(NO_3)_2$ needed to reach the equivalence point?

    (A) It would be cut in half.
    (B) It would decrease by a factor of 1.5.
    (C) It would double.
    (D) It would not change.

9. **D** The equivalence point is reached when there are equal moles of each reactant present. Diluting the sodium carbonate solution will not change the amount of it that is present in solution.

10.
$$2\ CO\ (g) + O_2\ (g) \rightarrow 2\ CO_2\ (g)$$

2.0 mol of CO (g) and 2.0 mol of O_2 (g) are pumped into a rigid, evacuated 4.0-L container, where they react to form CO_2 (g). Which of the following values does NOT represent a potential set of concentrations for each gas at a given point during the reaction?

	CO	O_2	CO_2
(A)	0.5	0.5	0
(B)	0	0.25	0.5
(C)	0.25	0.25	0.5
(D)	0.25	0.38	0.25

10. **C** For every two moles of CO that reacts, one mole of O_2 will also react. If one mole of CO reacts, half a mole of O_2 will react. At that point, the concentration of CO will be (1.0 mol/4.0 L = 0.25 M) and the concentration of the O_2 will be (1.5 mol/4.0 L = 0.38 M), rendering choice C impossible.

QUESTIONS	EXPLANATIONS

11. Neutral atoms of chlorine are bombarded by high-energy photons, causing the ejection of electrons from the various filled subshells. Electrons originally from which subshell would have the highest velocity after being ejected?

   (A) 1s
   (B) 2p
   (C) 3p
   (D) 3d

11. **C** The further away an electron is from the nucleus, the less binding energy that the incoming photons need to overcome, and as a result, the more kinetic energy the electron will have after it is ejected. The 3p subshell is the furthest one that a neutral chlorine atom would have electrons in.

12. A sample of oxygen gas at 50 °C is heated, reaching a final temperature of 100 °C. Which statement best describes the behavior of the gas molecules?

   (A) Their velocity increases by a factor of two.
   (B) Their velocity increases by a factor of four.
   (C) Their kinetic energy increases by a factor of 2.
   (D) Their kinetic energy increases by a factor of less than 2.

12. **D** Temperature in Kelvins is a measure of kinetic energy. When that temperature doubles, so does the kinetic energy. However, a doubling of the temperature in Celsius, while increasing the amount of kinetic energy, would not double it.

13. The average mass, in grams, of one mole of carbon atoms is equal to:

   (A) The average mass of a single carbon atom, measured in amus.
   (B) The ratio of the number of carbon atoms to the mass of a single carbon atom.
   (C) The number of carbon atoms in one amu of carbon.
   (D) The mass, in grams, of the most abundant isotope of carbon.

13. **A** This is a straightforward concept to understand. The average atomic mass of an element on the Periodic Table measures both the average mass of a mole of atoms of that element in grams, as well as being the average mass of a single atom of that element in amus.

14.

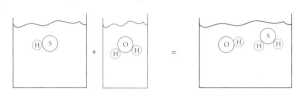

The diagram above best represents which type of reaction?

(A) Acid/base
(B) Oxidation/reduction
(C) Precipitation
(D) Decomposition

14. **A** A proton is transferred from the water to the HS⁻ ion, making the reaction an acid/base reaction.

15. Which of the following is true for all bases?

(A) All bases donate OH⁻ ions into solution.
(B) Only strong bases create solutions in which OH⁻ ions are present.
(C) Only strong bases are good conductors when dissolved in solution.
(D) For weak bases, the concentration of the base exceeds the concentration of OH⁻ ions in solution.

15. **C** Strong bases dissociate to donate hydroxide ions into a solution and weak bases cause the creation of hydroxide ions by taking protons from water. However, weak do not protonate strongly, meaning there will not be nearly as many ions in solution for a weak base as there are for a strong base. Thus, weak bases are not good conductors.

Use the following information to answer questions 16-18.

$14 H^+ (aq) + Cr_2O_7^{2-} (aq) + 3 Ni (s) \rightarrow 2 Cr^{3+} (aq) + 3 Ni^{2+} (aq) + 7 H_2O (l)$

In the above reaction, a piece of solid nickel is added to a solution of potassium dichromate.

16. Which species is being oxidized and which is being reduced?

	Oxidized	Reduced
(A)	$Cr_2O_7^{2-}$ (aq)	Ni (s)
(B)	Cr^{3+} (aq)	Ni^{2+} (aq)
(C)	Ni (s)	$Cr_2O_7^{2-}$ (aq)
(D)	Ni^{2+} (aq)	Cr^{3+} (aq)

16. **C** The $Cr_2O_7^{2-}$ (aq) gains electrons to change the oxidation state on Cr from +6 to +3. The Ni (s) loses electrons to change the oxidation state on Ni from 0 to +2.

QUESTIONS	EXPLANATIONS

17. How many moles of electrons are transferred when 1 mole of potassium dichromate is mixed with 3 mol of nickel?

   (A) 2 moles of electrons
   (B) 3 moles of electrons
   (C) 5 moles of electrons
   (D) 6 moles of electrons

17. **D** In one mole of dichromate, two moles of chromium are reduced, each requiring three moles of electrons. In three mole of nickel, three moles of nickel are being oxidized, reach requiring two moles of electrons.

18. How would the pH of the solution change as the reaction progresses?

   (A) It increases until the solution becomes basic.
   (B) It increases, but the solution remains acidic.
   (C) It decreases until the solution becomes basic.
   (D) It decreases, but the solution remains acidic.

18. **B** As the reaction progress, the H^+ ions are converting to H_2O molecules. This would raise the pH of the solution. As the reaction progresses, the solution will become less acidic, but there is no mechanism for it to become basic.

19. A sample of an unknown chloride compound was dissolved in water, and then titrated with excess $Pb(NO_3)_2$ to create a precipitate. After drying, it is determined there are 0.0050 mol of precipitate present. What mass of chloride is present in the original sample?

   (A) 0.177 g
   (B) 0.355 g
   (C) 0.522 g
   (D) 0.710 g

19. **B** The precipitate is $PbCl_2$, and in every mole of precipitate there are two moles of chlorine ions. Thus, there are 0.010 mol of chlorine present in the precipitate, and through conservation of mass, the same number of moles of chlorine were present in the original sample. 0.010 mol $\times$ 35.5 g = 0.355 g

QUESTIONS	EXPLANATIONS

20. Valence electrons in which of the atoms represented below would have the highest amount of potential energy relative to the nucleus?

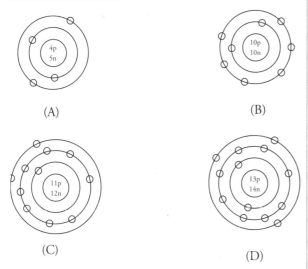

(A)

(B)

(C)

(D)

20. **B** The closer the valence electrons are to the nucleus, the more Coulombic potential energy they have. Though both have two shells, the atom in B (Ne) is smaller than the atom in A (Be) due to the extra protons, and thus the electrons will be closer to the nucleus.

21. The bond length between any two nonmetal atoms is achieved under which of the following distances?

(A) Where the energy of interaction between the atoms is at its minimum value

(B) Where the nuclei of each atom exhibits the strongest attraction to the electrons of the other atom

(C) The point at which the attractive and repulsive forces between the two atoms are equal

(D) The closest point at which a valence electron from one atom can transfer to the other atom

21. **A** The bond length always corresponds to the point where the potential bond energy (a balance of the attraction and repulsion forces between the two atoms) is at it's minimum value.

22. Hydrogen fluoride, HF, is a liquid at 15 °C. All other hydrogen halides (represented by HX, where X is any other halogen) are gases at the same temperature. Why?

(A) Fluorine has a very high electronegativity, therefore the H-F bond is stronger than any other H-X bond.

(B) HF is smaller than any other H-X molecule, therefore it exhibits stronger London dispersion forces.

(C) The dipoles in a HF molecule exhibit a particularly strong attraction force to the dipoles in other HF molecules.

(D) The H-F bond is the most ionic in character compared to all other hydrogen halides.

22. **C** The bonds between HF molecules are hydrogen bonds, which are very strong dipoles created due to the high electronegativity value of fluorine. No other hydrogen halide exhibits hydrogen bonding, and thus they would have weaker IMFs (and lower boiling points) than HF.

23.

	Initial pH	pH after NaOH addition
Acid 1	3.0	3.5
Acid 2	3.0	5.0

Two different acids with identical pH are placed in separate beakers. Identical portions of NaOH are added to each beaker, and the resulting pH is indicated in the table above. What can be determined about the strength of each acid?

(A) Acid 1 is a strong acid and acid 2 is a weak acid because acid 1 resists change in pH more effectively.

(B) Acid 1 is a strong acid and acid 2 is a weak acid because the NaOH is more effective at neutralizing the acid 2.

(C) Acid 1 is a weak acid and acid 2 is a strong acid because the NaOH must accept electrons from a larger percentage of acid 1 in order to neutralize it.

(D) Acid 1 is a weak acid and acid 2 is a strong acid because the concentration of the hydrogen ions will be greater in acid 2 after the NaOH addition.

23. C Weak acids resist changes in pH more effectively than strong atoms because so many molecules of weak acid are undissociated in solution. The base must cause those molecules to dissociate before affecting the pH significantly.

24. A stock solution of 12.0 M sulfuric acid is made available. What is the best procedure to make up 100. mL of 4.0 M sulfuric acid using the stock solution and water prior to mixing?

(A) Add 33.3 mL of water to then flask, then add 66.7 mL of 12.0 M acid.

(B) Add 33.3 mL of 12.0 M acid to then flask, then dilute it with 66.7 mL of water.

(C) Add 67.7 mL of 12.0 M acid to the flask, then dilute it with 33.3 mL of water.

(D) Add 67.7 mL of water to the flask, then add 33.3 mL of 12.0 M acid.

24. D Using $M_1V_1 = M_2V_2$, we can calculate the amount of stock solution needed. $(12.0)V_1 = (4.0)(100.0)$ $V_1 = 33.3$ mL. When making solutions, always add acid to water (in order to more effectively absorb the heat of the exothermic reaction).

QUESTIONS	EXPLANATIONS

Use the following data to answer questions 25-29

The enthalpy values for the several reactions are as follows:

(I) $C(s) + H_2O(g) \rightarrow CH_4(g) + H_2(g)$
$\Delta H = 131$ kJ/mol$_{rxn}$

(II) $CH_4(g) + H_2O(g) \rightarrow 3 H_2(g) + CO(g)$
$\Delta H = 206$ kJ/mol$_{rxn}$

(III) $CO(g) + H_2O(g) \rightarrow CO_2(g) + H_2(g)$
$\Delta H = -41$ kJ/mol$_{rxn}$

(IV) $CH_4(g) + 2 O_2(g) \rightarrow CO_2(g) + H_2O(l)$
$\Delta H = -890$ kJ/mol$_{rxn}$

25. In which of the the reactions does the energy of the bonds broken exceed the energy of the bonds formed by the greatest amount?

(A) Reaction I
(B) Reaction II
(C) Reaction III
(D) Reaction IV

25. **B** When calculating enthalpy, the total energy is always the bonds broken (reactants) minus the bonds formed (products). The more positive this value is, the more energy there is in the reactants compared to the products.

26. In which of the reactions is the value for ΔS the most positive?

(A) Reaction I
(B) Reaction II
(C) Reaction III
(D) Reaction IV

26. **A** The largest entropy changes arise from phase changes. Converting a solid into a gas greatly increases the disorder in the system, and thus the entropy.

27. Regarding reaction I, how would the addition of a catalyst effect the enthalpy and entropy changes for this reaction?

	Enthalpy	Entropy
(A)	Decrease	Decrease
(B)	Decrease	No Change
(C)	No Change	Decrease
(D)	No Change	No Change

27. **D** The addition of a catalyst speeds up the reaction, but would have no effect on neither the enthalpy or the entropy values.

28.

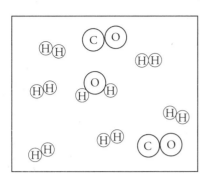

Regarding reaction II, to achieve the products present in the above diagram how many moles of each reactant must be present prior to the reaction?

(A) 1.0 mol of CH_4 and 2.0 mol of H_2O
(B) 2.0 mol of CH_4 and 2.0 mol of H_2O
(C) 2.0 mol of CH_4 and 3.0 mol of H_2O
(D) 3.0 mol of CH_4 and 2.0 mol of H_2O

28. **C** 2.0 mol of CH_4 would react with 2.0 moles of H_2O, leaving 1.0 mole left. It would also create 6.0 moles of H_2 and 2.0 moles of CO.

29. Regarding reaction IV, how much heat is absorbed or released when 2.0 mol of CH_4 (g) reacts with 2.0 mol of O_2 (g) ?

(A) 890 kJ of heat is released.
(B) 890 kJ of heat is absorbed.
(C) 1780 kJ of heat is released.
(D) 1780 kJ of heat is absorbed.

29. **A** The oxygen gas is limiting, and 2.0 moles of it will produce (negative signs mean exothermic reactions in which heat is released) exactly 890 kJ of heat.

30. London dispersion forces are caused by

(A) temporary dipoles created by the position of electrons around the nuclei in a molecule.
(B) the three-dimensional intermolecular bonding present in all covalent substances.
(C) the uneven electron-to-proton ratio found on individual atoms of a molecule.
(D) the electronegativity differences between the different atoms in a molecule.

30. **A** London dispersion forces are created by temporary dipoles due to the constant motion of electrons in an atom or molecule.

QUESTIONS	EXPLANATIONS

31. What is the general relationship between temperature and entropy for diatomic substances?

(A) They are completely independent of each other, temperature has no effect on entropy

(B) There is a direct relationship, as at higher temperatures there is an increase in energy dispersal

(C) There is an inverse relationship, as at higher temperatures substances are more likely to be in a gaseous state

(D) It depends on the specific substance and the strength of the intermolecular forces between individual molecules

31. **B** The higher the temperature of any substance, the larger the range of velocities the molecules of that substance can have, and thus the more disorder the substance can have. A Maxwell-Boltzmann diagram represents this in a graphical form.

32. Which of the following pairs of ions would make the best buffer with a basic pH?

(K_a for $HC_3H_2O_2 = 1.75 \times 10^{-5}$ and K_a for $HPO_4^{2-} = 4.8 \times 10^{-13}$)

(A) H_2SO_4 and Na_2SO_4
(B) HPO_4^{2-} and $H_2PO_4^-$
(C) $HC_3H_2O_2$ and $NaC_3H_2O_2$
(D) LiOH and NaOH

32. **B** Strong acids and bases do not make good buffers. To create a basic buffer, a base (such as HPO_4^{2-}, determined based on its K_a value) and its conjugate are needed.

33. A solution contains a mixture of four different compounds: KCl (*aq*), Fe(NO$_3$)$_3$ (*aq*), MgSO$_4$ (*aq*), and N$_2$H$_4$ (*aq*) Which of these compounds would be easiest to separate via distillation?

(A) KCl (*aq*)
(B) Fe(NO$_3$)$_3$ (*aq*)
(C) MgSO$_4$ (*aq*)
(D) N$_2$H$_4$ (*aq*)

33. **D** Distillation involves the boiling off of substances with different boiling points. The other three compounds are all ionic, meaning that in solution they are free ions. If we were to boil off all of the N$_2$H$_4$ and the water, the remaining ions would all mix together to form multiple precipitates. However, by heating the solution to a boiling point greater than that of N$_2$H$_4$ and lower than that of water, the N$_2$H$_4$ can be collected in a separate flask.

34.

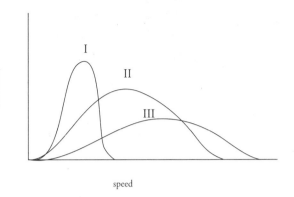

Identify the three gases represented on the Maxwell-Boltzmann diagram above. Assume all gases are at the same temperature.

	I	II	III
(A)	H_2	N_2	F_2
(B)	H_2	F_2	N_2
(C)	F_2	N_2	H_2
(D)	N_2	F_2	H_2

35. A sample of solid $MgCl_2$ would be most soluble in which of the following solutions?

(A) LiOH (*aq*)
(B) CBr_4 (*aq*)
(C) $Mg(NO_3)_2$ (*aq*)
(D) $AlCl_3$ (*aq*)

34. **C** At identical temperatures, the gases would all have identical amounts of kinetic energy. In order for that to happen, the gas with the lowest mass (H_2) would have to have the highest average velocity, and the gas with the highest mass (F_2) would have to have the lowest average velocity.

35. **A** Like dissolves like, so ionic substances dissolve best in ionic compounds, eliminating (B). Both (C) and (D) share an ion with $MgCl_2$, and via the common ion effect/Le Chateliers principle that would reduce the solubility of the $MgCl_2$ in those solutions.

QUESTIONS	EXPLANATIONS

36. Most transition metals share a common oxidation state of +2. Which of the following best explains why?

(A) Transition metals all have a minimum of two unpaired electrons.
(B) Transition metals have unstable configurations and are very reactive.
(C) Transition metals tend to gain electrons when reacting with other elements.
(D) Transition metals will lose their outermost s-block electrons when forming bonds.

36. **D** The outermost *s*-block electrons in a transition metal tend to be lost before the *d*-block eectrons do. Additionally, the other options do not accurately describe the properties of transition metals.

37. $2\ Ag^{+}\ (aq) + Fe\ (s) \rightarrow 2\ Ag\ (s) + Fe^{2+}\ (aq)$

Which of the following would cause an increase in potential in the voltaic cell described by the above reaction?

(A) Increasing $[Fe^{2+}]$
(B) Adding more Fe (*s*)
(C) Decreasing $[Fe^{2+}]$
(D) Removing some Fe (*s*)

37. **C** Adding or removing a solid would not cause any equilibrium shift. Decreasing the concentration of the Fe^{2+} causes a shift to the right, which would increase the potential of the cell.

QUESTIONS	EXPLANATIONS

Use the following information to answer questions 38-41.

Consider the Lewis structures for the following molecules:

$$CO_2, CO_3^{2-}, NO_2^-, \text{ and } NO_3^-$$

38. Which molecule would have the shortest bonds?

    (A) CO_2
    (B) CO_3^{2-}
    (C) NO_2^-
    (D) NO_3^-

39. Which molecules are best represented by multiple resonance structures?

    (A) CO_2 and CO_3^{2-}
    (B) NO_2^- and NO_3^-
    (C) CO_3^{2-} and NO_3^-
    (D) CO_3^{2-}, NO_2^- and NO_3^-

40. Which molecule or molecules exhibit sp^2 hybridization around the central carbon atom?

    (A) CO_2 and CO_3^{2-}
    (B) NO_2^- and NO_3^-
    (C) CO_3^{2-} and NO_3^-
    (D) CO_3^{2-}, NO_2^- and NO_3^-

The following Lewis structures are necessary to answer questions 38-41:

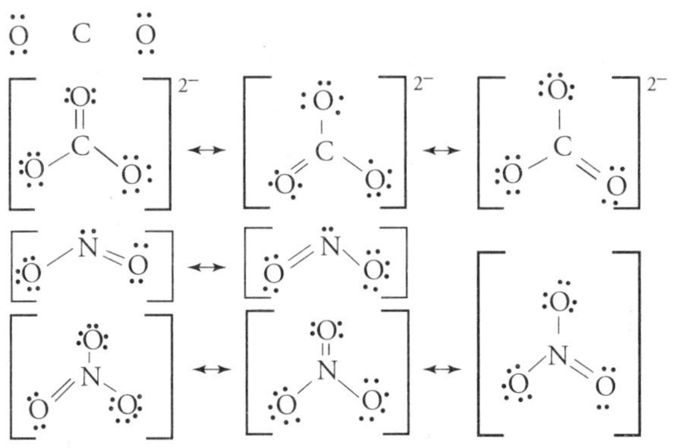

38. **A** CO_2 has a bond order of 2, which exceeds the order of the other strucutres.

39. **D** Except for CO_2, the other three molecules all display resonance.

40. **D** All three of those structures have three electron domains, and thus sp^2 hybridization.

41. Which molecule would have the smallest bond angle between terminal atoms?

    (A) CO_2
    (B) CO_3^{2-}
    (C) NO_2^-
    (D) NO_3^-

41. **C** The bond angle of NO_2^- would be less than that of NO_3^- or CO_3^{2-} because the unbonded pair of electrons reduces the overall bond angle.

42. $$NH_4^+(aq) + NO_2^-(aq) \rightarrow N_2(g) + 2\,H_2O(l)$$

    Increasing the temperature of the above reaction will increase the rate of reaction.

    Which of the following is NOT a reason that increased temperature increases reaction rate?

    (A) The reactants will be more likely to overcome the activation energy.
    (B) The number of collisions between reactant molecules will increase.
    (C) A higher distribution of reactant molecules will have high velocities.
    (D) Alternate reaction pathways become available at higher temperatures.

42. **D** Even if this were true, any pathways that become available at higher temperatures would be less likely to be taken than the original pathway.

43. Which of the following diagrams best represents what is happening on a molecular level when NaCl dissolves in water?

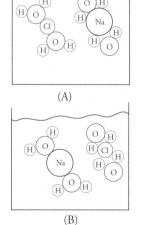

(A)

(B)

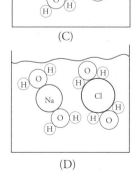

(C)

(D)

43. **D** The Na^+ cations would be attracted to the negative (oxygen) end of the water molecules and the Cl^- anions would be attracted to the positive (hydrogen) end. Additionally, Na^+ ions are smaller than Cl^- ions as they have less filled energy levels.

QUESTIONS	EXPLANATIONS

44. Nitrous acid, HNO_2, has a pK_a value of 3.3. If a solution of nitric acid is found to have a pH of 4.2, what can be said about the concentration of the conjugate acid/base pair found in solution?

    (A) $[HNO_2] > [NO_2^-]$
    (B) $[NO_2^-] > [HNO_2]$
    (C) $[H_2NO_2^+] > [HNO_2]$
    (D) $[HNO_2] > [H_2NO_2^+]$

44. **B** In this case, the pH is greater than the pK_a. This means that there will be more conjugate base present in solution than the original acid. The conjugate base of HNO_2 is NO_2^-.

45. Which of the following processes is an irreversible reaction?

    (A) $CH_4\ (g) + O_2\ (g) \rightarrow CO_2\ (g) + H_2O\ (l)$
    (B) $HCN\ (aq) + H_2O\ (l) \rightarrow CN^-\ (aq) + H_3O^+\ (aq)$
    (C) $Al(NO_3)_3\ (aq) \rightarrow Al^{3+}\ (aq) + 3\ NO_3^-\ (aq)$
    (D) $2\ Ag^+\ (aq) + Ti\ (s) \rightarrow 2\ Ag\ (s) + Ti^{2+}\ (aq)$

45. **A** Combustion reactions are irreversible, while the other reactions are examples of acid-base, dissolution, and oxidation-reduction, all of which are reversible in this case.

Use the following information to answer questions 46-50.

A sample of H_2S gas is placed in an evacuated, sealed container and heated until the following decomposition reaction occurs at 1000 K:

$$2\ H_2S\ (g) \rightarrow 2\ H_2\ (g) + S_2\ (g) \qquad K_c = 1.0 \times 10^{-6}$$

46. Which of the following represents the equilibrium constant for this reaction?

    (A) $K_c = \dfrac{[H_2]^2 [S_2]}{[H_2S]^2}$

    (B) $K_c = \dfrac{[H_2S]^2}{[H_2]^2 [S_2]}$

    (C) $K_c = \dfrac{2[H_2][S_2]}{2[H_2S]}$

    (D) $K_c = \dfrac{2[H_2S]}{2[H_2][S_2]}$

46. **A** Equilibrium is always products over reactants, and coefficients in a balanced equilibrium reaction become exponents in the equilibrium expression.

QUESTIONS	EXPLANATIONS

47. Which of the following graphs would best represent the change in concentration of the various species involved in the reaction over time?

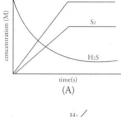
(A)

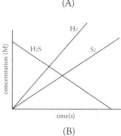
(B)

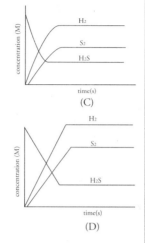
(C)

(D)

47. **C** The concentration of the H_2S will decrease exponentially until it reaches a constant value. The concentrations of the two products will increase exponentially (the H_2 twice as quickly as the S_2) until reaching equilibrium.

48. Which option best describes what will happen to the reaction rates if the pressure on the system is increased after it has reached equilibrium?

(A) The rate of both the forward and reverse reactions will increase.
(B) The rate of the forward reaction will increase while the rate of the reverse reaction decreases.
(C) The rate of the forward reaction will decrease while the rate of the reverse reaction increases.
(D) Neither the rate of the forward nor reverse reactions will change.

48. **C** Increasing the pressure causes a shift towards the side with less gas molecules—in this case, a shift to the left. This means the reverse reaction rate increases while the forward reaction rate decreases until equilibrium is re-established.

QUESTIONS	EXPLANATIONS
49. If, at a given point in the reaction, the value for the reaction quotient Q is determined to be 2.5×10^{-8}, which of the following is occurring?	49. **A** The reaction will progress until $Q = K_c$. If $Q < K_c$, the numerator of the expression (the products) will continue to increase while the denominator (the reactant) increases until equilibrium is established.

49. If, at a given point in the reaction, the value for the reaction quotient Q is determined to be 2.5×10^{-8}, which of the following is occurring?

(A) The concentration of the reactant is decreasing while the concentration of the products is increasing.

(B) The concentration of the reactant is increasing while the concentration of the products is decreasing.

(C) The system has passed the equilibrium point and the concentration of all species involved in the reaction will remain constant.

(D) The concentrations of all species involved are changing at the same rate.

49. **A** The reaction will progress until $Q = K_c$. If $Q < K_c$, the numerator of the expression (the products) will continue to increase while the denominator (the reactant) increases until equilibrium is established.

50. As the reaction progresses at a constant temperature of 1000K, how does the value for the Gibbs free energy constant for the reaction change?

(A) It stays constant.
(B) It increases exponentially.
(C) It increases linearly.
(D) It decreases exponentially.

50. **A** If the temperature is constant, then the equilibrium constant K is unchanged. Via $G = -RT \ln K$, if K and T are both constant then so is the value for ΔG.

51. An unknown substance is found to have a high melting point. In addition, it is a poor conductor of electricity and does not dissolve in water. The substance most likely contains

(A) ionic bonding.
(B) nonpolar covalent bonding.
(C) covalent network bonding.
(D) metallic bonding.

51. **C** An ionic substance would dissolve in water, and a nonpolar covlaent substance would have a low melting point. A metallic substance would be a good conductor. The only type of bonding that meets all the criteria is covalent network bonding.

52. Which of the following best explains why the ionization of atoms can occur during photoelectron spectroscopy, even though ionization is not a thermodynamically favored process?

(A) It is an exothermic process due to the release of energy as an electron is liberated from the Coulombic attraction holding it to the nucleus.

(B) The entropy of the system increases due to the separation of the electron from its atom.

(C) Energy contained in the light can be used to overcome the Coulombic attraction between electrons and the nucleus.

(D) The products of the ionization are at a lower energy state than the reactants.

52. **C** Light contains energy (via $E = h\nu$), and that energy can be used to cause a reaction.

53.

$$2H_2O_2 \, (aq) \longrightarrow 2H_2O \, (l) + O_2 \, (g)$$

H–O–O–H (x8)	H–O–O–H (x4) H–O–H (x4) O=O (x2)	H–O–O–H (x2) H–O–H (x6) O=O (x3)
$t = 0s$	$t = 200s$	$t = 400s$

The above diagrams shows the decomposition of hydrogen peroxide in a sealed container in the presence of a catalyst. What is the overall order for the reaction?

(A) Zero order
(B) First order
(C) Second order
(D) Third order

53. **B** In 200 seconds, half of the original sample decayed. In another 200s, half of the remaining sample decayed. This demonstrates a first order reaction.

54.

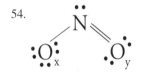

One of the resonance structures for the nitrite ion is shown above. What is the formal charge on each atom?

	O_x	N	O_y
(A)	–1	+1	–1
(B)	+1	–1	0
(C)	0	0	–1
(D)	–1	0	0

54. **D** O_x has 6 valence electrons and 7 assigned electrons: $6 - 7 = -1$. Both O_y and the N atoms have the same number of valence and assigned electrons, making their formal charges zero.

QUESTIONS	EXPLANATIONS
Use the following information to answer questions 55-57.	55. **B** The strength of an atom's magnetic moment increases with an increase in the number of unpaired electrons. Nitrogen has the most unpaired electrons (3), and thus the strongest magnetic moment.

Use the following information to answer questions 55-57.

Atoms of four elements are examined: carbon, nitrogen, neon, and sulfur.

55. Atoms of which element would have the strongest magnetic moment?

   (A) carbon
   (B) nitrogen
   (C) neon
   (D) sulfur

55. **B** The strength of an atom's magnetic moment increases with an increase in the number of unpaired electrons. Nitrogen has the most unpaired electrons (3), and thus the strongest magnetic moment.

56. Atoms of which element are most likely to form a structure with the formula XF_6 (where X is one of the four atoms)

   (A) carbon
   (B) nitrogen
   (C) neon
   (D) sulfur

56. **D** Sulfur is the only element with an empty d-block in its outermost energy level, and is thus the only atom of the four which can form an expanded octet.

57. Which element would have a photoelectron spectra in which the peak representing electrons with the lowest ionization energy would be a three times higher than all other peaks?

   (A) carbon
   (B) nitrogen
   (C) neon
   (D) sulfur

57. **C** Neon has six electrons in its subshell with the lowest ionization energy ($2p$), and only two electrons in the other two subshells ($1s$ and $2s$).

58. The above diagram supports which of the following conclusions about the reaction shown below?

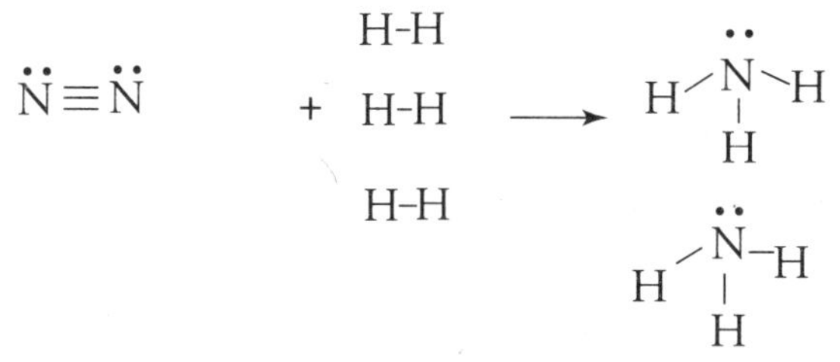

$$\ddot{N} \equiv \ddot{N} \quad + \quad \begin{matrix} H\text{-}H \\ H\text{-}H \\ H\text{-}H \end{matrix} \quad \longrightarrow \quad \begin{matrix} H \diagup \overset{\cdot\cdot}{N} \diagdown H \\ \underset{H}{|} \\ \\ H \diagup \overset{\cdot\cdot}{N} \diagup H \\ \underset{H}{|} \end{matrix}$$

   (A) There is an increase in entropy.
   (B) Mass is conserved in all chemical reactions.
   (C) The bond length increases after the reaction.
   (D) The enthalpy value is positive.

58. **B** The amount of matter is equal on both sides of the reaction. None of the other options are supported by the diagram.

---

59. $NO_2 + O_3 \rightarrow NO_3 + O_2$ slow

     $NO_3 + NO_2 \rightarrow N_2O_5$ fast

A proposed reaction mechanism for the reaction of nitrogen dioxide and ozone is detailed above. Which of the following is the rate law for the reaction?

   (A) rate $= k[NO_2][O_3]$
   (B) rate $= k[NO_3][NO_2]$
   (C) rate $= k[NO_2]^2[O_3]$
   (D) rate $= k[NO_3][O_2]$

59. **A** The overall rate law is always equal to the rate law of the slowest elementary step. The rate law of any elementary step can be determined using the coefficients of the reactants in that step.

60.

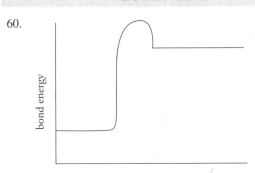

The concentrations of the reactants and products in the reaction represented by the above coordinate are found are found to be changing very slowly. Which of the following statements best describes the reaction a given that the reaction is exergonic? ($\Delta G < 0$)

(A) The reaction is under kinetic control.
(B) The reaction has reached a state of equilibrium.
(C) The reaction is highly exothermic in nature.
(D) The addition of heat will increase the rate of reaction significantly.

60. **A** Reactions with high activation energies that do not proceed at a measurable rate are considered to be under kinetic control—that is, their rate of progress is based on kinetics instead of thermodynamics.

1. A stock solution of 0.100 M cobalt (II) chloride is used to create several solutions, indicated in the data table below:

Sample	Volume $CoCl_2$ (mL)	Volume H_2O (mL)
1	20.00	0
2	15.00	5.00
3	10.00	10.00
4	5.00	15.00

(a) In order to achieve the degree of accuracy shown in the table above select which of the following pieces of laboratory equipment could be used when measuring out the $CoCl_2$:

150-mL beaker
400-mL beaker
250-mL Earlenmeyer flask
50-mL buret
50-mL graduated cylinder
100-mL graduated cylinder

(a) The 50-mL buret is what is needed here in order to get measurements that are accurate to the hundreths place. Graduated cylinders are generally accurate to the tenths, and using flasks or beakers to measure out volume is highly inaccurate.

(b) Calculate the concentration of the $CoCl_2$ in each sample. The solutions are then placed in cuvettes before being inserted into a spectrophotometer calibrated to 560 nm and their values are measured, yielding the data below:

Sample	Absorbance
1	0.485
2	0.364
3	0.243
4	0.121

(b) Sample 1: 0.100 M (no dilution occurred)

For samples 2-4, the first step is to calculate the number of moles of $CoCl_2$, using the Molarity = moles/volume formula. Then, divide the number of moles by the new volume (volume of $CoCl_2$ + volume H_2O) to determine the new concentration.

Sample 2: 0.100 M = n/0.015 L
n = 0.0015 mol
0.0015 mol/0.020 L = 0.075 M

Sample 3: 0.100 M = n/0.010 L
n= 0.0010 mol
0.0010 mol/0.020 L = 0.050 M

Sample 4: 0.100 M = n/0.0050 L
n = 0.00050 mol
0.00050 mol/0.202 L = 0.025 M

QUESTIONS	EXPLANATIONS

(c) If gloves are not worn when handling the cuvettes, how might this affect the absorbance values gathered?

(c) If gloves are not worn when handling the cuvettes, fingertip oils may be deposited on the surface of the cuvette. These oils may absorb some light, increasing the overall absorbance values.

(d) If the path length of the cuvette is 1.00 cm, what is the molar absorptivity value for $CoCl_2$ at 560 nm?

(d) Using Beer's Law: $A = abc$ and the data from Sample 1:

$0.485 = a(1.00 \text{ cm})(0.100 \text{ M})$ $a = 4.85 \text{ M}^{-1}\text{cm}^{-1}$

Values from any sample can be used with identical results.

(e) On the axes below, plot a graph of absorbance vs. concentration. The y-axes scale is set, and be sure to scale the x-axes appropriately.

(e)

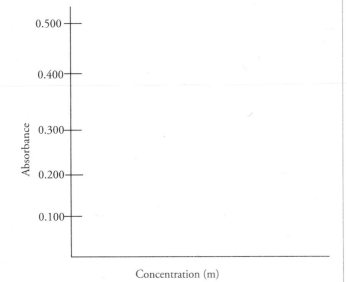

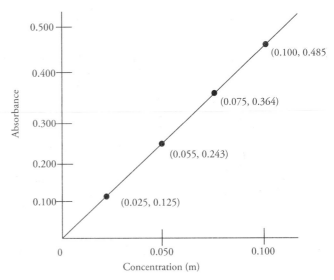

QUESTIONS	EXPLANATIONS

QUESTIONS

(f) What would the absorbance values be for $CoCl_2$ solutions at the following concentrations?

   (i) 0.067 M

   (ii) 0.180 M

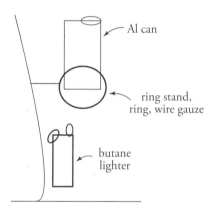

- Al can
- ring stand, ring, wire gauze
- butane lighter

2. A sample of liquid butane (C_4H_{10}) in a pressurized lighter is set up directly beneath an aluminum can, as show in the diagram above. The can contains 100.0 mL of water, and when the butane is ignited the temperature of the water inside the can increases from 25.0 °C to 82.3 °C. The total mass of butane ignited is found to be 0.51 g, the specific heat of water is 4.18 J/g °C, and the density of water is 1.00 g/mL.

  (a) Write the balanced chemical equation for the combustion of butane in air.

EXPLANATIONS

(f) While the absorbance values for (I) can be estimated from a correctly-drawn graph, the absorbance value for (ii) would be off the chart. The more accurate method to determine the absorbances of both samples would be to determine the slope of the line from the graph in (e) and then plug in the appropriate values. (Note: Estimating the value for (i) is still an acceptable solution.)

Any two points can be taken to determine the slope of the graph, but by examining the Beer's Law calculation in (d) we can see it is already in slope-intercept form (as the y-intercept would be zero, as at 0 concentration there would be no absorbance). So, the slope of the line is equal to $4.85 \text{ M}^{-1}\text{cm}^{-1} \times 1.00 \text{ cm} = 4.85 \text{ M}^{-1}$

(i) $A = (4.85 \text{ M}^{-1})(0.067 \text{ M}) = 0.325$

(ii) $A = (4.85 \text{ M}^{-1})(0.180 \text{ M}) = 0.873$

(a) The butane reacts with the oxygen in the air, and the products of any hydrocarbon combustion are always carbon dioxide and water.
$2 \text{ C}_4\text{H}_{10}(g) + 9 \text{ O}_2(g) \rightarrow 8 \text{ CO}_2(g) + 10 \text{ H}_2\text{O}(g)$

QUESTIONS	EXPLANATIONS
(b) (i) How much heat did the water gain?	(i) The formula needed is $q = mc\Delta T$. The mass in the equation is the mass of the water, which is equal to 100.0 g (as the density of water is 1.0 g/mL). $\Delta T = 82.3\ °C - 25.0\ °C = 57.3\ °C$ and c is given as 4.18 J/g°C so: $q = (100.0\ g)(4.18\ J/g°C)(57.3\ °C) = 24{,}000\ J$ or 24 kJ
(ii) What is the experimentally determined heat of combustion for butane based on this experiment? Your answer should be in kJ/mol.	(ii) 0.51 g $C_4H_{10} \times$ 1 mol C_4H_{10}/58.14 g C_4H_{10} = 0.0088 mol C_4H_{10} 24 kJ/0.0088 mol C_4H_{10} = 2,700 kJ/mol
(c) Given butane's density of 0.573 g/mL at 25 °C, calculate how much heat would be emitted if 5.00 mL of it were combusted at that temperature.	(c) 0.573 g/mL = m/5.00 mL m = 2.87 g butane 2.87 g $C_4H_{10} \times$ 1 mol C_4H_{10}/58.14 g C_4H_{10} = 0.049 mol C_4H_{10} 0.049 mol C_4H_{10} * 2700 kJ/mol = 130 kJ of heat emitted
(d) The overall combustion of butane is an exothermic reaction. Explain why this is in terms of bond energies.	(d) The overall reaction is exothermic because the bond energy of the reactants exceeds that of the products.
(e) One of the major sources of error in this experiment comes from the heat which is absorbed by the air. Why, then, might it not be a good idea to perform this experiment inside a sealed container to prevent the heat from leaving the system?	(e) Performing this reaction in a sealed container would limit the amount of oxygen available to react. To get an accurate heat of combustion for butane, it needs to be the limiting reactant, and if the oxygen runs out prior to all the butane combusting that will not happen. Alternatively, the pressure of the sealed container could exceed the strength of the container and cause an explosion. There are approximately twice as many gas molecules after the reaction than there are prior to the reaction, and that means the pressure would approximately double if the reaction goes to completion.

QUESTIONS	EXPLANATIONS

3.

$$2 N_2O_5 (g) \rightarrow 4 NO_2 (g) + O_2 (g)$$

The data below was gathered for the decomposition of N_2O_5 at 310 K via the equation above.

Time (s)	$[N_2O_5]$ (M)
0	0.250
500.	0.190
1000.	0.145
2000.	0.085

(a) How does the rate of appearance of NO_2 compare to the rate of disappearance of N_2O_5? Justify your answer.

(b) The reaction is determined to be first order overall. On the axes below, create a graph of some function of concentration vs. time that will produce a straight line. Label and scale your axes appropriately.

(a) Due to the stoichimetric ratios, 4 moles of NO_2 are created for every 2 moles of N_2O_5 that decomposes. Therefore, the rate of appearance of NO_2 will be twice the rate of disappearance for N_2O_5. As that rate is constantly changing over the course of the reaction, it is impossible to get exact values, but the ratio of 2:1 will stay constant.

(b) First order decompositions always create a straight line when plotting the natural log of concentration of the reactant vs. time.

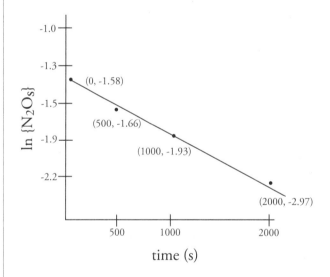

QUESTIONS	EXPLANATIONS
(c) (i) What is the rate constant for this reaction? Include units!	(i) Interpreting the graph using slope-intercept form, we get $\ln[N_2O_5]_t = -kt + \ln[N_2O_5]_0$ Any (non-intercept) values on the graph can be plugged in to determine the rate constant. Using at $[N_2O_5] = 0.190$ M at $t = 500$s: $\ln(0.190) = -k(500.) + \ln(0.250)$ $-1.66 = -500.\ k + -1.39$ $-.27 = -500.\ k$ $k = 5.40 \times 10^{-4}$ In terms of units, if rate $= k[N_2O_5]$, then via analyzing the units: $M/s = k(M) \qquad k = s^{-1}$ So $k = 5.40 \times 10^{-4}$ s^{-1}. Any point along the line will give the same value (as it is equal to the negative slope of the line).
(ii) What would the concentration of N_2O_5 be at $t = 1500$ s?	(ii) Using the equation from c(i): $\ln[N_2O_5]_{1500} = -5.40 \times 10^{-4}$ s$^{-1}(1500.\ \text{s}) + \ln(0.250)$ $\ln[N_2O_5]_{1500} = -2.20$ $[N_2O_5]_{1500} = 0.111$ M
(iii) What is the half-life of N_2O_5?	(iii) The half life is defined as how long it takes half of the original sample to decay. Thus, at the half-life of N_2O_5 for this reaction, $[N_2O_5] = 0.125$ M. So: $\ln(0.125) = -5.40 \times 10^{-4}(t) + \ln(0.250)$ $-2.08 = -5.40 \times 10^{-4}(t) + -1.39$ $-0.69 = -5.40 \times 10^{-4}(t)$ $t = 1280$ s

QUESTIONS	**EXPLANATIONS**

(d) Would the addition of a catalyst increase, decrease, or have no effect on the following variables? Justify your answers.:

(i) Rate of disappearance of N_2O_5

(ii) Magnitude of the rate constant

(iii) Half-life of N_2O_5

(i) Catalysts speed up the reaction, so adding a cataylst would increase the rate of disappearance of N_2O_5.

(ii) If the reaction is moving faster, the slope of any line graphing concentration vs. time would be steeper. This increases the magnitude of the slope, and thus, increases the magnitude of the rate constant as well.

(iii) If the reaction is proceeding at a faster rate, it will take less time for the N_2O_5 to decompose, and thus the half-life of the N_2O_5 would decrease.

4. Consider the Lewis structures for the following four molecules:

n-Butylamine

Propanal

Pentane

Methanol

(a) All of the substances are liquids at room temperature. Organize them from high to low in terms of boiling points, clearly differentiating between the intermolecular forces in each substance.

(a) The strongest type of intermolecular force present here is hydrogen bonding, which both n-Butylamine and methanol exhibit. Of the two, n-Butylamine would have the stronger London disperion forces as it has more electrons (is more polarizable), and so it has stronger IMFs than methanol. For the remaining two structures, propanal has permanent dipoles while pentane is completely nonpolar. Therefore, propanal would have stronger IMFs than pentane.

So, from high to low boiling point: n-Butylamine > methanol > propanal > pentane

QUESTIONS	**EXPLANATIONS**

(b) On the methanol diagram reproduced below, draw the locations of all dipoles.

(b)

$$H-C \overset{\delta^+}{-} \overset{\delta^-}{O}-\overset{\delta^+}{H}$$

(c) n-Butylamine is found to have the lowest vapor pressure at room temperature out of the four liquids. Justify this observation in terms of intermolecular forces.

(c) Vapor pressure arises from molecules overcoming the IMFs to other molecules and escaping the surface of the liquid to become a gas.. Due to its hydrogen bonding and large size, n-Butylamine would have the strongest IMFs and thus its molecules would have the hardest time escaping the surface, leading to a low vapor pressure.

5. Current is run through a solution of aqueous solution of 1.0 M nickel (II) chloride, and a gas is evolved at the right-hand electrode, as indicated by the diagram below:

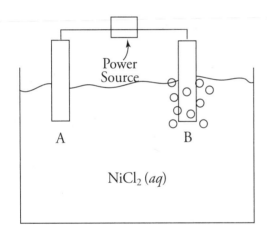

Power Source

A   B

NiCl$_2$ (aq)

The standard reduction potential for several reactions is given in the following table:

Half-cell	E^o_{red}
Cl$_2$ (g) + 2e$^-$ → 2 Cl$^-$	+1.36 V
O$_2$ (g) + 4 H$^+$ + 4e$^-$ → 2 H$_2$O (l)	+1.23 V
Ni^{2+} + 2e$^-$ → Ni (s)	–0.25 V
2 H$_2$O (l) + 2e$^-$ → H$_2$ (g) + 2 OH$^-$	–0.83 V

(a) Determine which half-reaction is occurring at each electrode:
   (i) Oxidation:

(i) The two choices for oxidation reactions are
2 Cl$^-$ → Cl$_2$ (g) + 2e$^-$ (because chloride ions are present in the solution initially) and
2 H$_2$O (l) → O$_2$ (g) + 4 H$^+$ + 4 e$^-$ (because oxidation is defined as the loss of electrons). As we flipped the reactions we must also flip the signs, so for the chloride reaction $E^o_{ox} = -1.36$ V and for the water one $E^o_{ox} = -1.23$ V. The half-reaction with the positive sign is more likely to occur, so the answer is
2 H$_2$O (l) → O$_2$ (g) + 4 H$^+$ + 4 e$^-$

QUESTIONS	EXPLANATIONS

QUESTIONS

(ii) Reduction:

(b)

   (i) Calculate the standard cell potential for the cell.

   (ii) Calculate the Gibbs free energy value for the cell.

(c) Which electrode in the diagram (A or B) is the cathode, and which is the anode? Justify your answers.

EXPLANATIONS

(ii) Using the same logic as above, the two choices for reduction reaction are $Ni^{2+} + 2 e^-$ Ni (s) and $2 H_2O$ $(l) + 2 e^- \rightarrow H_2(g) + 2 OH^-$. Unlike the oxidation reaction, there is no need to flip the signs on these half-reactions. The more positive value in this case belongs to the nickel reaction, so the answer is $Ni^{2+} + 2 e^- \rightarrow Ni$ (s)

(i) $E^o_{cell} = E^o_{ox} + E^o_{red} = $ -1.23 V + (-0.25 V) = –0.98 V

(ii) $\Delta G = nFE$

n is equal to 4 moles of electrons (from the oxidation reaction). Even though $n = 2$ in the unbalanced reduction reaction, to balance the reaction the reduction half-reaction would need to be multiplied by 2 so the electrons balance. Additionally, a volt is equal to a Joule/Coulomb, which is what we will use to get our units to make sense. So:

$\Delta G = $ –(4 mol e$^-$)(96,500 C/mol e$^-$)(–0.98 J/C)

$\Delta G = $ 380,000 J or 380 kJ

(c) Oxygen gas is evolved in the oxidation reaction, meaning that is occuring at electrode B. Oxidation always occurs at the anode, so electrode A is the cathode and electrode B is the anode.

QUESTIONS	EXPLANATIONS

6. Aniline, $C_6H_5NH_2$ is a weak base with $K_b = 3.8 \times 10^{-10}$.

   (a)   Write out the reaction which occurs when aniline reacts with water.

(a)   The aniline will act as a proton acceptor and take a proton from the water.

$$C_6H_5NH_2\,(aq) + H_2O\,(l) \leftrightarrow C_6H_5NH_3^+\,(aq) + OH^-\,(aq)$$

   (b)   (i) What is the concentration of each species at equilibrium in a solution of 0.25 M $C_6H_5NH_2$?

(b) (i) For the above equilibrium, $K_b = [C_6H_5NH_3^+][OH^-] / [C_6H_5NH_2]$. The concentrations of both the conjugate acid and the hydroxide ion will be equal, and the concentration of the aniline itself will be approximately the same, as it is a weak base which has a very low protonation rate. You can do an ICE chart to confirm this, but it is not really necessary if you understand the concepts underlying weak acids and bases.

$3.8 \times 10^{-10} = (x)(x)/(0.25)$
$x^2 = 9.5 \times 10^{-11}$
$x = 9.7 \times 10^{-6}$

$[C_6H_5NH_3^+] = [OH^-] = 9.7 \times 10^{-6}\,M$ and
$[C_6H_5NH_2] = 0.25$ M

        (ii) What is the pH value for the solution in b(i)?

(ii) $pOH = -\log[OH^-]$
$pOH = -\log(9.7 \times 10^{-6}\,M)$
$pOH = 5.01$

$pOH + pH = 14$
$5.01 + pH = 14$
$pH = 8.99$

7. A rigid, sealed 12.00 L container is filled with 10.00 g each of three different gases: CO_2, NO, and NH_3. The temperature of the gases is held constant 35.0 °C. Assume ideal behavior for all gases.

   (a)   (i) What is the mole fraction of each gas?

(i)   First, the moles of each gas needs calculating:

CO_2 = 10.00 g × 1 mol/44.01 g = 0.227 mol
NO = 10.00 g × 1 mol/30.01 g = 0.333 mol
NH_3 = 10.00g × 1 mol/17.03 g = 0.587 mol

Total moles of gas = 1.147 moles

X_{CO2} = 0.227/1.147 = 0.198
X_{NO} = 0.333/1.147 = 0.290
X_{NH3} = 0.587/1.147 = 0.512

QUESTIONS	EXPLANATIONS

(ii) What is the partial pressure of each gas?

(ii) Using the ideal gas law, we can calculate the total pressure in the container:

$$PV = nRT$$

$$P\,(12.00\text{ L}) = (1.147\text{ mol})(0.0821\text{ atmL/molK})(308\text{ K})$$

$$P = 2.42\text{ atm}$$

The partial pressure of a gas is equal to the total pressure time the mole fraction of that gas.

$$P_{CO2} = (2.42\text{ atm})(0.198) = 0.479\text{ atm}$$
$$P_{NO} = (2.42\text{ atm})(0.290) = 0.702\text{ atm}$$
$$P_{NH3} = (2.42\text{ atm})(0.512) = 1.24\text{ atm}$$

(b) Out of the three gases, molecules of which gas will have the highest velocity? Why?

(b) If all gases are at the same temperature, they have the same amount of kinetic energy. KE has aspects of both mass and velocity, so the gas with the lowest mass would have the highest velocity. Thus, the NH_3 molecules have the highest velocity.

(c) Name one circumstance in which the gases might deviate from ideal behavior, and clearly explain the reason for the deviation.

(c) The most common reason for deviation from ideal behavior is when the intermolecular forces of the gas molecules act upon each other. This would occur when the molecules are very close together and/or moving very slowly, so deviations would occur at high pressures and/or low temperatures.

Completely darken bubbles with a No. 2 pencil. If you make a mistake, be sure to erase mark completely. Erase all stray marks.

1.

YOUR NAME: _____
(Print)   Last   First   M.I.

SIGNATURE: _____   DATE: __ / __ / __

HOME ADDRESS: _____
(Print)   Number and Street

_____
City   State   Zip Code

PHONE NO.: _____

IMPORTANT: Please fill in these boxes exactly as shown on the back cover of your test book.

2. TEST FORM

3. TEST CODE

0	A	J	0	0
1	B	K	1	1
2	C	L	2	2
3	D	M	3	3
4	E	N	4	4
5	F	O	5	5
6	G	P	6	6
7	H	Q	7	7
8	I	R	8	8
9			9	9

4. REGISTRATION NUMBER

0	0	0	0	0	0	0
1	1	1	1	1	1	1
2	2	2	2	2	2	2
3	3	3	3	3	3	3
4	4	4	4	4	4	4
5	5	5	5	5	5	5
6	6	6	6	6	6	6
7	7	7	7	7	7	7
8	8	8	8	8	8	8
9	9	9	9	9	9	9

6. DATE OF BIRTH

Month	Day		Year	
○ JAN				
○ FEB	0	0	0	0
○ MAR	1	1	1	1
○ APR	2	2	2	2
○ MAY	3	3	3	3
○ JUN		4	4	4
○ JUL		5	5	5
○ AUG		6	6	6
○ SEP		7	7	7
○ OCT		8	8	8
○ NOV		9	9	9
○ DEC				

7. GENDER
○ MALE
○ FEMALE

5. YOUR NAME

First 4 letters of last name				FIRST INIT	MID INIT
A	A	A	A	A	A
B	B	B	B	B	B
C	C	C	C	C	C
D	D	D	D	D	D
E	E	E	E	E	E
F	F	F	F	F	F
G	G	G	G	G	G
H	H	H	H	H	H
I	I	I	I	I	I
J	J	J	J	J	J
K	K	K	K	K	K
L	L	L	L	L	L
M	M	M	M	M	M
N	N	N	N	N	N
O	O	O	O	O	O
P	P	P	P	P	P
Q	Q	Q	Q	Q	Q
R	R	R	R	R	R
S	S	S	S	S	S
T	T	T	T	T	T
U	U	U	U	U	U
V	V	V	V	V	V
W	W	W	W	W	W
X	X	X	X	X	X
Y	Y	Y	Y	Y	Y
Z	Z	Z	Z	Z	Z

The Princeton Review®

1. A B C D
2. A B C D
3. A B C D
4. A B C D
5. A B C D
6. A B C D
7. A B C D
8. A B C D
9. A B C D
10. A B C D
11. A B C D
12. A B C D
13. A B C D
14. A B C D
15. A B C D
16. A B C D
17. A B C D
18. A B C D
19. A B C D
20. A B C D

21. A B C D
22. A B C D
23. A B C D
24. A B C D
25. A B C D
26. A B C D
27. A B C D
28. A B C D
29. A B C D
30. A B C D
31. A B C D
32. A B C D
33. A B C D
34. A B C D
35. A B C D
36. A B C D
37. A B C D
38. A B C D
39. A B C D
40. A B C D

41. A B C D
42. A B C D
43. A B C D
44. A B C D
45. A B C D
46. A B C D
47. A B C D
48. A B C D
49. A B C D
50. A B C D
51. A B C D
52. A B C D
53. A B C D
54. A B C D
55. A B C D
56. A B C D
57. A B C D
58. A B C D
59. A B C D
60. A B C D

Completely darken bubbles with a No. 2 pencil. If you make a mistake, be sure to erase mark completely. Erase all stray marks.

1.

YOUR NAME: _____
(Print)

Last   First   M.I.

SIGNATURE: _____   DATE: ___ / ___ / ___

HOME ADDRESS: _____
(Print)

Number and Street

_____
City   State   Zip Code

PHONE NO.: _____

IMPORTANT: Please fill in these boxes exactly as shown on the back cover of your test book.

2. TEST FORM

_____

3. TEST CODE   **4. REGISTRATION NUMBER**

⓪	Ⓐ	Ⓙ	⓪	⓪	⓪	⓪	⓪	⓪	⓪	⓪	⓪
①	Ⓑ	Ⓚ	①	①	①	①	①	①	①	①	①
②	Ⓒ	Ⓛ	②	②	②	②	②	②	②	②	②
③	Ⓓ	Ⓜ	③	③	③	③	③	③	③	③	③
④	Ⓔ	Ⓝ	④	④	④	④	④	④	④	④	④
⑤	Ⓕ	Ⓞ	⑤	⑤	⑤	⑤	⑤	⑤	⑤	⑤	⑤
⑥	Ⓖ	Ⓟ	⑥	⑥	⑥	⑥	⑥	⑥	⑥	⑥	⑥
⑦	Ⓗ	Ⓠ	⑦	⑦	⑦	⑦	⑦	⑦	⑦	⑦	⑦
⑧	Ⓘ	Ⓡ	⑧	⑧	⑧	⑧	⑧	⑧	⑧	⑧	⑧
⑨			⑨	⑨	⑨	⑨	⑨	⑨	⑨	⑨	⑨

6. DATE OF BIRTH

Month	Day		Year	
◯ JAN				
◯ FEB	⓪	⓪	⓪	⓪
◯ MAR	①	①	①	①
◯ APR	②	②	②	②
◯ MAY	③	③	③	③
◯ JUN		④	④	④
◯ JUL		⑤	⑤	⑤
◯ AUG		⑥	⑥	⑥
◯ SEP		⑦	⑦	⑦
◯ OCT		⑧	⑧	⑧
◯ NOV		⑨	⑨	⑨
◯ DEC				

7. GENDER
◯ MALE
◯ FEMALE

5. YOUR NAME

First 4 letters of last name				FIRST INIT	MID INIT
Ⓐ	Ⓐ	Ⓐ	Ⓐ	Ⓐ	Ⓐ
Ⓑ	Ⓑ	Ⓑ	Ⓑ	Ⓑ	Ⓑ
Ⓒ	Ⓒ	Ⓒ	Ⓒ	Ⓒ	Ⓒ
Ⓓ	Ⓓ	Ⓓ	Ⓓ	Ⓓ	Ⓓ
Ⓔ	Ⓔ	Ⓔ	Ⓔ	Ⓔ	Ⓔ
Ⓕ	Ⓕ	Ⓕ	Ⓕ	Ⓕ	Ⓕ
Ⓖ	Ⓖ	Ⓖ	Ⓖ	Ⓖ	Ⓖ
Ⓗ	Ⓗ	Ⓗ	Ⓗ	Ⓗ	Ⓗ
Ⓘ	Ⓘ	Ⓘ	Ⓘ	Ⓘ	Ⓘ
Ⓙ	Ⓙ	Ⓙ	Ⓙ	Ⓙ	Ⓙ
Ⓚ	Ⓚ	Ⓚ	Ⓚ	Ⓚ	Ⓚ
Ⓛ	Ⓛ	Ⓛ	Ⓛ	Ⓛ	Ⓛ
Ⓜ	Ⓜ	Ⓜ	Ⓜ	Ⓜ	Ⓜ
Ⓝ	Ⓝ	Ⓝ	Ⓝ	Ⓝ	Ⓝ
Ⓞ	Ⓞ	Ⓞ	Ⓞ	Ⓞ	Ⓞ
Ⓟ	Ⓟ	Ⓟ	Ⓟ	Ⓟ	Ⓟ
Ⓠ	Ⓠ	Ⓠ	Ⓠ	Ⓠ	Ⓠ
Ⓡ	Ⓡ	Ⓡ	Ⓡ	Ⓡ	Ⓡ
Ⓢ	Ⓢ	Ⓢ	Ⓢ	Ⓢ	Ⓢ
Ⓣ	Ⓣ	Ⓣ	Ⓣ	Ⓣ	Ⓣ
Ⓤ	Ⓤ	Ⓤ	Ⓤ	Ⓤ	Ⓤ
Ⓥ	Ⓥ	Ⓥ	Ⓥ	Ⓥ	Ⓥ
Ⓦ	Ⓦ	Ⓦ	Ⓦ	Ⓦ	Ⓦ
Ⓧ	Ⓧ	Ⓧ	Ⓧ	Ⓧ	Ⓧ
Ⓨ	Ⓨ	Ⓨ	Ⓨ	Ⓨ	Ⓨ
Ⓩ	Ⓩ	Ⓩ	Ⓩ	Ⓩ	Ⓩ

The Princeton Review

1. Ⓐ Ⓑ Ⓒ Ⓓ
2. Ⓐ Ⓑ Ⓒ Ⓓ
3. Ⓐ Ⓑ Ⓒ Ⓓ
4. Ⓐ Ⓑ Ⓒ Ⓓ
5. Ⓐ Ⓑ Ⓒ Ⓓ
6. Ⓐ Ⓑ Ⓒ Ⓓ
7. Ⓐ Ⓑ Ⓒ Ⓓ
8. Ⓐ Ⓑ Ⓒ Ⓓ
9. Ⓐ Ⓑ Ⓒ Ⓓ
10. Ⓐ Ⓑ Ⓒ Ⓓ
11. Ⓐ Ⓑ Ⓒ Ⓓ
12. Ⓐ Ⓑ Ⓒ Ⓓ
13. Ⓐ Ⓑ Ⓒ Ⓓ
14. Ⓐ Ⓑ Ⓒ Ⓓ
15. Ⓐ Ⓑ Ⓒ Ⓓ
16. Ⓐ Ⓑ Ⓒ Ⓓ
17. Ⓐ Ⓑ Ⓒ Ⓓ
18. Ⓐ Ⓑ Ⓒ Ⓓ
19. Ⓐ Ⓑ Ⓒ Ⓓ
20. Ⓐ Ⓑ Ⓒ Ⓓ

21. Ⓐ Ⓑ Ⓒ Ⓓ
22. Ⓐ Ⓑ Ⓒ Ⓓ
23. Ⓐ Ⓑ Ⓒ Ⓓ
24. Ⓐ Ⓑ Ⓒ Ⓓ
25. Ⓐ Ⓑ Ⓒ Ⓓ
26. Ⓐ Ⓑ Ⓒ Ⓓ
27. Ⓐ Ⓑ Ⓒ Ⓓ
28. Ⓐ Ⓑ Ⓒ Ⓓ
29. Ⓐ Ⓑ Ⓒ Ⓓ
30. Ⓐ Ⓑ Ⓒ Ⓓ
31. Ⓐ Ⓑ Ⓒ Ⓓ
32. Ⓐ Ⓑ Ⓒ Ⓓ
33. Ⓐ Ⓑ Ⓒ Ⓓ
34. Ⓐ Ⓑ Ⓒ Ⓓ
35. Ⓐ Ⓑ Ⓒ Ⓓ
36. Ⓐ Ⓑ Ⓒ Ⓓ
37. Ⓐ Ⓑ Ⓒ Ⓓ
38. Ⓐ Ⓑ Ⓒ Ⓓ
39. Ⓐ Ⓑ Ⓒ Ⓓ
40. Ⓐ Ⓑ Ⓒ Ⓓ

41. Ⓐ Ⓑ Ⓒ Ⓓ
42. Ⓐ Ⓑ Ⓒ Ⓓ
43. Ⓐ Ⓑ Ⓒ Ⓓ
44. Ⓐ Ⓑ Ⓒ Ⓓ
45. Ⓐ Ⓑ Ⓒ Ⓓ
46. Ⓐ Ⓑ Ⓒ Ⓓ
47. Ⓐ Ⓑ Ⓒ Ⓓ
48. Ⓐ Ⓑ Ⓒ Ⓓ
49. Ⓐ Ⓑ Ⓒ Ⓓ
50. Ⓐ Ⓑ Ⓒ Ⓓ
51. Ⓐ Ⓑ Ⓒ Ⓓ
52. Ⓐ Ⓑ Ⓒ Ⓓ
53. Ⓐ Ⓑ Ⓒ Ⓓ
54. Ⓐ Ⓑ Ⓒ Ⓓ
55. Ⓐ Ⓑ Ⓒ Ⓓ
56. Ⓐ Ⓑ Ⓒ Ⓓ
57. Ⓐ Ⓑ Ⓒ Ⓓ
58. Ⓐ Ⓑ Ⓒ Ⓓ
59. Ⓐ Ⓑ Ⓒ Ⓓ
60. Ⓐ Ⓑ Ⓒ Ⓓ

NOTES

NOTES

NOTES

Score Your Best on Both of the Tests

Get proven prep for the SAT and ACT at **PrincetonReviewBooks.com**

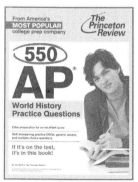

Ace the APs:

Cracking the AP Biology Exam, 2014 Edition
978-0-8041-2410-2
$18.99/$21.95 Can.
eBook: 978-0-8041-2411-9

Cracking the AP Calculus AB & BC Exams, 2014 Edition
978-0-307-94618-8
$19.99/$22.95 Can.
eBook: 978-0-307-94619-5

Cracking the AP Chemistry Exam, 2014 Edition
978-0-307-94620-1
$18.99/$21.95 Can.
eBook: 978-0-307-94621-8

Cracking the AP Economics Macro & Micro Exams, 2014 Edition
978-0-8041-2412-6
$18.00/$20.00 Can.
eBook: 978-0-8041-2413-3

Cracking the AP English Language & Composition Exam, 2014 Edition
978-0-8041-2414-0
$18.00/$20.00 Can.
eBook: 978-0-8041-2415-7

Cracking the AP English Literature & Composition Exam, 2014 Edition
978-0-8041-2416-4
$18.00/$20.00 Can.
eBook: 978-0-8041-2417-1

Cracking the AP Environmental Science Exam, 2014 Edition
978-0-8041-2418-8
$18.99/$21.95 Can.
eBook: 978-0-8041-2419-5

Cracking the AP European History Exam, 2014 Edition
978-0-307-94622-5
$18.99/$21.95 Can.
eBook: 978-0-307-94623-2

Cracking the AP Human Geography Exam, 2014 Edition
978-0-8041-2420-1
$18.00/$20.00 Can.
eBook: 978-0-8041-2421-8

Cracking the AP Physics B Exam, 2014 Edition
978-0-8041-2422-5
$18.99/$21.95 Can.
eBook: 978-0-8041-2423-2

Cracking the AP Physics C Exam, 2014 Edition
978-0-8041-2424-9
$18.99/$21.95 Can.
eBook: 978-0-8041-2425-6

Cracking the AP Psychology Exam, 2014 Edition
978-0-8041-2426-3
$18.00/$20.00 Can.
eBook: 978-0-8041-2427-0

Cracking the AP Spanish Exam with Audio CD, 2014 Edition
978-0-8041-2428-7
$24.99/$27.95 Can.

Cracking the AP Statistics Exam, 2014 Edition
978-0-8041-2429-4
$19.99/$22.95 Can.
eBook: 978-0-8041-2430-0

Cracking the AP U.S. Government & Politics Exam, 2014 Edition
978-0-8041-2431-7
$18.99/$21.95 Can.
eBook: 978-0-8041-2432-4

Cracking the AP U.S. History Exam, 2014 Edition
978-0-307-94624-9
$18.99/$21.95 Can.
eBook: 978-0-307-94625-6

Cracking the AP World History Exam, 2014 Edition
978-0-307-94626-3
$19.99/$22.95 Can.
eBook: 978-0-307-94627-0

Extra AP Practice & Prep:

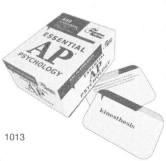

550 AP Calculus AB & BC Practice Questions
978-0-8041-2445-4
$16.99/$18.95 Can.
eBook: 978-0-8041-2446-1

550 AP U.S. Government & Politics Practice Questions
978-0-8041-2443-0
$16.99/$18.95 Can.
eBook:
978-0-8041-2444-7

550 AP World History Practice Questions
978-0-8041-2441-6
$16.99/$18.95 Can.
eBook: 978-0-8041-2442-3

550 AP Biology Practice Questions *August 2014!*
978-0-8041-2488-1
$16.99/19.99 Can.

550 AP European History Practice Questions *August 2014!*
978-0-8041-2490-4
$16.99/19.99 Can.

Essential AP Biology (Flashcards)
978-0-375-42803-6
$18.99/$20.99 Can.

Essential AP Psychology (Flashcards)
978-0-375-42801-2
$18.99/$20.99 Can.

Essential AP U.S. Government & Politics (Flashcards)
978-0-375-42804-3
$18.99/$20.99 Can.

Essential AP U.S. History (Flashcards)
978-0-375-42800-5
$18.99/$20.99 Can.

Essential AP World History (Flashcards)
978-0-375-42802-9
$18.99/$20.99 Can.

Available everywhere books are sold and at PrincetonReviewBooks.com.